# Regulating Social Media

D1501262

# COMMUNICATION LAW

Susan Drucker
*General Editor*

Vol. 2

The Communication Law series
is part of the Peter Lang Media and Communication list.
Every volume is peer reviewed and meets
the highest quality standards for content and production.

PETER LANG
New York • Washington, D.C./Baltimore • Bern
Frankfurt • Berlin • Brussels • Vienna • Oxford

# Regulating Social Media

## Legal and Ethical Considerations

EDITED BY **Susan J. Drucker** AND **Gary Gumpert**

PETER LANG
New York • Washington, D.C./Baltimore • Bern
Frankfurt • Berlin • Brussels • Vienna • Oxford

**Library of Congress Cataloging-in-Publication Data**

Regulating social media: legal and ethical considerations /
edited by Susan J. Drucker, Gary Gumpert.
pages cm. — (Communication law; vol. 2)
Includes bibliographical references and index.
1. Social media—Law and legislation—United States.
I. Drucker, Susan J., editor of compilation.
II. Gumpert, Gary, editor of compilation.
KF390.5.C6.R44    343.7309'99—dc23    2012042558
ISBN 978-1-4331-1484-7 (hardcover)
ISBN 978-1-4331-1483-0 (paperback)
ISBN 978-1-4539-0972-0 (e-book)
ISSN 2153-1390


Bibliographic information published by **Die Deutsche Nationalbibliothek**.
**Die Deutsche Nationalbibliothek** lists this publication in the "Deutsche
Nationalbibliografie"; detailed bibliographic data is available
on the Internet at http://dnb.d-nb.de/.

The paper in this book meets the guidelines for permanence and durability
of the Committee on Production Guidelines for Book Longevity
of the Council of Library Resources.

© 2013 Peter Lang Publishing, Inc., New York
29 Broadway, 18th floor, New York, NY 10006
www.peterlang.com

Printed in the United States of America

# CONTENTS

# Preface

A preface is somewhat of an anomaly as it is an introductory or preliminary statement to a work, but it is probably always written at the end after the authors have learned about what they intended to do and perhaps did . . . or did not. As Alice in Wonderland discovered, it's confusing to think backwards, but any book is confusing in conception because the end has first to be discovered before the beginning can be written.

A volume composed of splendid and valued contributors is a challenge sometimes difficult to complete as an overall forward-looking vision often has to be reconsidered as each chapter assumes a personality of its own and provides insights and data not considered in the beginning.

Many, in frustration, have noted that the Law always plays catch-up and Ethics is by definition, a reaction to behavior and conduct and thus a murky, but rich setting, but with the addition of Social Media, the playground gets a bit overcrowded. The combination is pushed into new vistas with the constant acceleration of Social Media—a communal collective pushed to the edges by changing and convergent communication technologies.

To attempt to keep up with technological innovation in the communication world is to assume the role of the "cockeyed optimist," because it simply can't be done in the understanding sense. The acceleration of technology is bewildering and the pace extraordinary. One of the Drucker-Gumpert team is reaching "four score" years—born into a world without television, computer, Internet, CD, or DVD and more. In those days our communication consciousness consisted of telegraph, radio, wired recorders, 78-rpm records, and newspapers. Over thirty years ago that co-editor wrote of "media grammar and generation gaps" in which the thesis was espoused that each of us is shaped, in part, by the technology into which we emerge from the womb; that the media environment defines our generation. We now live among the Twitter generation with a consciousness of 140-characters able to connect with virtually anyone in the world equipped with a multi-functional mobile device and a short-term memory but accompanied with a global curiosity. What are the implications of "I am therefore I Tweet?"

This media generation has nimble fingers and the ability to make a smart phone sing. When faced with technological frustration we seek anyone below the age of 12 who will miraculously know what to do. We marvel at the sense of direction of these digital natives until we realize they are, at all times, "GPS guided." . . . lost without a technological mentor or companion. Our closest relationships are reshaped by social media encounters. The engagement is a public event negotiated on-line. Pregnancies are discovered; births announced. Family feuds fueled by Facebook postings. As we go to press over 1 billion users have joined Facebook.

In short we are in a state of constant redefinition interacting with a ceaseless shifting technological environment. This volume attempts to address the complex interaction of ethics, law, and social media. We hope that we have raised fundamental questions that will remain relevant at least for several years.

It is customary to thank those innocent individuals who helped us assemble this volume and, of course, we are grateful to them and their insightful analysis. One of us also suggested that we thank our mentors beginning with Aristotle and Plato, but we desisted going in that direction because it would make the Preface too long. We thought of thanking all of our "Facebook Friends." But that would have insulted our "Linkedin" colleagues. We could thank those patient souls who have helped us enter and navigate the social media world, but we can do that online. We also thought of thanking our families, but they have suffered through those clichés before. The possibility also arose to thank our editor at Lang, but Mary Savigar and her colleagues have undergone such pledges of fealty before. We finally concluded that while there were too many to whom we owed too much, that we would thank our parents who allowed us to enter this tumultuous world in which there is too much to learn, but which we persist on seeking to understand without any hope in sight.

*Susan J. Drucker*
*Gary Gumpert*
*Great Neck, New York, Oct. 2012*

# · 1 ·

# Introduction

## Thoughts on Social Media, Law, and Ethics

Susan J. Drucker & Gary Gumpert

The introduction to any volume is a strange exercise generally conceived of after the body of work has been written. By that time, hopefully, the editor(s) or author(s) are aware of the scope and vision of the work. What has emerged in the case of *Regulating Social Media: Laws and Ethics* is the complexity and immensity of trying to define the term in question. In the spirit and hope of achieving some clarity, we began the task of definition after the completion of the book.

We would, in all probability, agree on three definitional statements.

1. All media of communication are social.
2. All social media are either public and/or private, although the distinction is increasingly blurred.
3. All social media are, by definition, regulated institutionally and/or by tacitly agreed upon cultural values.

All of this assumes that we know what a social medium is—that we recognize it when it is encountered and/or when it is imposed upon us. The definitions abound.

The Merriam-Webster online dictionary defines social media as "forms of electronic communication (as Web sites for social networking and microblogging) through which users create online communities to share information, ideas, personal messages, and other content (as videos)"(*Merriam-Webster Dictionary*, 2012). It attributes the first usage of the term to 2004.

Dictionary.com defines social media as "Web sites and other online means of communication that are used by large groups of people to share information and to develop social and professional contacts"(Dictionary.com, 2012).

Researcher danah m. boyd defines social media "as web-based services that allow individuals to (1) construct a public or semi-public profile within a bounded system, (2) articulate a list of other users with whom they share a connection, and (3) view and traverse their list of connections and those made by others within the system. The nature and nomenclature of these connections may vary from site to site" (boyd & Ellison, 2007).

It is curious that the current usage of the term *social media* is directly linked to the Internet, but certainly the degree to which the electronic world of social communication has grown is staggering. Blogger Jeff Bullas's (2012) figures are amazing:

1. One in every nine people on Earth is on Facebook. (This number is calculated by dividing the planet's 6.94 billion people by Facebook's 750 million users.)
2. People spend 700 billion minutes per month on Facebook.
3. Each Facebook user spends on average 15 hours and 33 minutes a month on the site.
4. More than 250 million people access Facebook through their mobile devices.
5. More than 2.5 million Web sites have integrated with Facebook.
6. 30 billion pieces of content are shared on Facebook each month.
7. 300,000 users helped translate Facebook into 70 languages.

Yet the definition of social media is surprisingly elusive. One of our industry contacts, when asked for a simple definition, stated the following: "The simplest definition of 'social media' is 'any form of communication that allows many members of a community to interact freely with each other to share observations, opinions, and other nonsense'" (Cohen, 2012).

Along this line of thought, we began to reflect upon appropriate criteria that defined in and defined out the nature of social media. In the pursuit of a family history on a totally unrelated project, we came across a set of postcards, one of which has a painting on it of three 19th-century ladies sitting on a village street knitting. On the other side was a note written by the mother of one of the editors of this volume, dated December 16, 1915, and addressed to her mother.

We began to wonder who else had read these words written almost 100 years ago. It suddenly occurred to us that this was an example of an early social medium.

The postcard emerged out of Austria and Germany in the mid-19th century and became a "social craze" by the beginning of the 20th century. The early postcard included room for a handwritten message and an address on one side with a painting and later a photograph on the reverse side. The selection of the postcard required the choice of an appropriate artistic expression plus a personal, carefully handwritten message. The art of the handwritten message, of penmanship, was a direct expression of one's persona. The postcard constituted an advertisement of the self—an interpersonal medium open to public inspection and perusal. The postcard was not simply a private means of communication but served as, and continues to be, albeit in a dwindling fashion, a public social medium. To some extent, the postcard has been driven into obsolescence by the newer technology. The iPhone camera documents each and every moment, transporting the self onto the other or the many others of Facebook.

This side excursion into the past helps define the scope of this volume—because any medium of communication defined by its publicness, its sharedness, invites, probably requires, some element of regulation or ethical monitoring. Even the postcard had to be regulated. The United States issued prestamped postal cards in 1873.

> The Unites States Postal Service was the only one allowed to print the cards until May 19, 1898 when Congress passed the Private Mailing Card Act which then allowed private firms to produce cards. The private mailing cards cost one cent to mail instead of the letter rate which was two cents. The term "Private Mailing Card" was required to be printed on cards that were not printed by the United States Postal Service. (*The History of Postcards*)

The more interesting question was how the content of postcards was regulated. Certainly, in addition to the operational requirements—placement of the address and space given to the message, the postal authorities were the arbiters of good taste. Some degree of regulation was necessary, but with the introduction of electronic technology communication, from the telegraph onward, communication had to be regulated in order to avoid traffic chaos. Concomitantly, independent of electronic traffic control, the more public the medium becomes, the more prevalent the ethical and moral issues become.

The matter of control of communication that is simultaneously private and public is fascinating. The postcard, in contrast to its relative, the sealed letter, consisted of a personal message imbedded in a visible public conveyance. In retrospect, the media of personal communication can be divided into private, pub-lic, and private-public variations. With the rise of digital communication and the Internet, this division of openness or privacy becomes rather complicated. Even the relatively simple e-mail becomes a convoluted phenomenon. Is it pri-vate or public? Is it private and secure or is it available to others for inspection? Does the individual have control of his or her privacy? On the other hand, when is Facebook private and when is it public? Is to be "friended" a public act or is "unfriending" a personal denial of existence? Social media represents a multi-faceted regulatory landscape influenced and determined by a mélange of legal, social, ethical, and often hidden dimensions.

Having said that, we return to our earlier observation that all media of com-munication—even those prior to the Internet—are and have been social. No medium is "un" or "not" social. All media bridge and connect, but the social nature of a medium can be and is limited by number and context.

It may be helpful to reflect upon three influential factors that have shaped and delineated the social medium milieu: convergence, community, and globalization.

Convergence refers to the merging nature of the Internet and digital tech-nology. Watching a television program on a computer redefines what is tele-vision. Using the iPad as a book redefines the process of reading and therefore the book. Previously thought of separate and discrete media become fused with each other, and the individual platform of dissemination then serves multiple functions, thereby redefining the platform itself as a unique multifunc-tional medium. Diverse definitions and conceptualizations of convergence are useful, with some addressing the content, some featuring the business or cor-porate mergers of media industries, and some highlighting interactivity of users. "Media convergence embraces both technological and organizational con-vergence" (Fishman, 2010, p. 124). In his volume, *Convergence Culture*, MIT Media Studies Professor Henry Jenkins (2006) notes that, "convergence is a word…that manages to describe technological, industrial, cultural, and social changes depending on who's speaking and what they think they are talking about" (pp. 2–3). He emphasizes resulting media trends such as unprecedented audience participation and the distribution of a single franchise through diverse forms of media. Digitalization is central to convergence. The functions once rel-egated to distinct silos now bleed across media. Print, photos, video, and sound are transported across the same channels and reside in the same media envi-ronments, rendering almost obsolete old distinctions long essential to regula-tory and ethical considerations. Social media support text, audio, video, and images accessible via personal computer, handheld device, smart phone, or

Internet-enabled television. Convergence dissolves the limits associated with media governance.

Traditionally, the limits of free expression evolved with due consideration to the nature of the media/medium to be regulated. Social media environments are classic exemplars of convergent media. Convergent media defy traditional categorization by fusing distinct media, each with its unique grammar, syntax, and conventions. The personal letter, radio, television, sound recordings, motion pictures, telephone, facsimile, etc. are now transmitted, received, experienced through common delivery services. Social Web sites are by definition converged media environments accessible via personal computer, handheld device television set and smart phone. The content is a mélange of text, audio, photograph, and video. Convergence involves inter-site linking. Yet old regulatory issues of transmission and content persist (Drucker, Gumpert, & Cohen, 2010, p. 67).

Increasingly, conventional mass media and interpersonal media environments merge into a single media environment facilitated by convergence, offering infinite digital media choices. Users have been empowered by digital media to select, demand, and filter at will, thereby transforming old media and patterns of use. Production and consumption are seamlessly reversed. Consumer production becomes the norm. Social media are inherently networking platforms that serve as forms of self-expression and self-promotion, allowing individuals to broadcast news about themselves. Tweeting assumes the need or wish to receive such information and in so doing, the individual achieves public persona or celebrity status. Ultimately, this egocentric worldview is cultivated. According to a study conducted by "Crowd Science" calculating measured attitudes toward social media, particularly Myspace, Facebook, and Twitter, the attraction of social media is the emphasis on "'me' that unites fa'me' and social 'me'dia" (Solis, 2011). Lives of selectivity reshape conceptions of and commitments to community.

Community implies relationships existing in one or more sites. With increasing dependence on mediated communication, social relationships once thought of requiring contiguity are redefined to transcend location. The community of place is potentially transformed into a community of places—individual and definable sites of relationships connected through a medium of communication. The elimination of place as a requirement of community and relationship is achievable through convergent technology. The nonpropinquitous community, one not defined by the physical notion of nearness, creates a supramobile mentality. "Me media" offer the choice of community based on similarity, common interests, values, and relevance.

Globalization accompanies spatial redefinition and adds the complexity of governance and administration. Some element of policy and regulation accompanies communication, transcending, independent of place. An integrated world economy, technologically connected, requires a regulated system in which connection transcends place and site. Globalization of media has often been cast as a world becoming a more integrated market, an age of global technical linkage, in which old concepts of media and cultural imperialism, homogenization, and hybridization are shaped by media technologies. But media and communication industries go beyond merely facilitating overall globalization. International media and global civil society have been credited with the new insurgencies of the Arab Spring of 2011 and other instances of political contention. Media are instrumental in the transformation of social, cultural, and political structures, sometimes opening cultures for consumerism, sometimes protest, sometimes self-promotion or selective interest-based connectivity, irrespective of physical boundaries. Globalization beyond markets, goods, and services initiates the exchange of ideas. Castells's model of a space of flows asserts that material and immaterial flows have created a new spatial logic that has overtaken the historically accepted "space of places" (Castells, 1989).

The physical parameters of space are no longer a constraint to connection. Mobile devices have ushered in an era of media consumption irrespective of space. Social networking by way of mobile devices is enabling connection to social media anywhere, anytime, spread throughout the day or continuously and obsessively. Over 1.3 billion users are expected to access social media from mobile devices by 2016, almost doubling the figures for 2011, according to a report published by Juniper Research (Agarwal, 2011). By mid-2011, more than 72 million Americans had accessed social networking sites or blogs via their mobile devices, a figure representing a 126% increase in one year (Van Grove, 2011). According to a study conducted by comScore, nearly one third of all U.S. mobile users now access social media services (i.e., 40 million Americans) on an almost daily basis. comScore reports online social networking via mobile devices is growing rapidly in Europe—a 44% increase in one year in the UK, France, Germany, Italy, and Spain (comScore, 2011). The regularity of access increased as well with a 67% growth in "almost daily" mobile access to social media, an increase largely attributable to the availability of apps (comScore, 2011). In Germany, a 2012 survey revealed that 77% of Internet users in that country access social media via a mobile device occasionally, with regular or heavy usage via mobile devices more prevalent among young adults and teens (McNaughton, 2012). The Juniper report noted the following:

The trend to integrate social, local and mobile experiences is driving the geosocial phenomena. People want to find out not only what their friends are doing, but also their location and other available activities in the area. Geosocial networks are particularly suited to the mobile space as most smartphones now include GPS, and have an "always on, always connected" experience. (Agarwal, 2011)

Mobile media feeds constant connection, a trend documented in "The World Unplugged Project," published by the International Center for Media & the Public Agenda (the world UNPLUGGED, 2012). The study documents behavior and reactions in 10 countries to the absence of mobile connection. The noted reactions include the following: "Fretful, Confused, Anxious, Irritable, Insecure, Nervous, Restless, Crazy, Addicted, Panicked, Jealous, Angry, Lonely, Dependent, Depressed, Jittery and Paranoid" (Alleyne, 2011).

## The Regulatory Environment

Social media space presents a staggering breadth of legal and ethical matters to consider: copyright and trademark, along with defamation, privacy, harassment, stalking, contracts, advertising, and censorship issues, to name a few. Myriad standards of professional ethics command compliance in order for various media industries to function.

Frequently, the terms *social media* and *social network* are used interchangeably, but there is a difference between the terms, which may be significant when considered from a regulatory perspective.

Adding to the complexities of the regulatory environment is a lack of clarity with regard to the nature of communication subject to governance or sanction. Social media is a way to transmit or share information with a broad audience. Users have the ability and opportunity to produce and disseminate content. Social networking is about engagement or association with individuals sharing, based on some common interest. Most social networks require mutual consent before members are considered "connected." Social networks feature relationships and community building. Social media, like other channels of communication, feature the delivery of information, whereas social networking foregrounds interaction and conversation. Social networking offers direct communication between those who choose to connect with each other and other like-minded people introduced in these environments. Social media is more akin to an outlet for broadcasting, while social networking is a utility for connecting with others.

LinkedIn is a social networking site while YouTube is a social media site. Facebook and Twitter are hybrids offering characteristics of both. While there are common characteristics between social media and social networking, they are not synonymous, and it is in the distinctions we find a source of regulatory nuisance and difference. Historically, the regulation of interpersonal interactional contexts is distinguished from mediated transmission/broadcast of messages to an audience. Free speech and free press have long been distinguished. In U.S. jurisprudence, Justice Potter Stewart has noted

> That the First Amendment speaks separately of freedom of speech and freedom of the press is no constitutional accident, but an acknowledgment of the critical role played by the press in American society. The Constitution requires sensitivity to that role, and to the special needs of the press in performing it effectively. (*Houchins v. KQED*, 1978).

The unique nature of each medium (qualitative and quantitative differences in the mode of transmission and audience) has been the basis of regulation of mediated communication. The differences between interactional/speech and dissemination/press can offer a helpful lens through which to unravel the thorny legal and ethical issues explored in this volume.

The legal landscape of social media varies widely, moving from social medium to social network to social bookmarking (sharing multiple links with stakeholders). There are different regulatory frames for dealing with blogging (narrative writing) and microblogging (providing real-time updates and obtaining feedback). Photo sharing and video sharing sites both can raise further distinct regulatory contexts.

The social media scene is evolving at a dizzyingly fast pace, so it is to be expected there will be vexing, perhaps disturbing, certainly novel, legal issues continually emerging. One truism of law is that the law always lags behind media innovations. This maxim can be extended to ethical principles as well. Technological and industry developments set in motion developing norms, rules, regulations, directives, litigation, and legislation. The tumultuous regulatory environment is revealed in old-fashioned news headlines and in tweets perhaps to be stored in the Cloud:

- "Internet and Social Network Expert in Iran Jailed" (New Knight Center, 2012).
- From Palestine: "Concern About Health Of Woman Journalist Held For Facebook Comments" (Concern About, 2012).
- "Phony Facebook Page Tests NJ Identity Theft Law" (Gottlieb, 2011).

- Reports of arrests come from Guatemala and Venezuela, with tweeting messages deemed a threat to the banking system in Brazil (Reporters Without Borders, 2012).
- U.S. employees have been fired for posting "inappropriate" information on Twitter, Facebook, and YouTube, respectively (Fired over Facebook, 2011).
- Employees have been fired for actively job hunting through LinkedIn. In the litigation that followed, the employer argued successfully that the employee criticized the company in public (New Knight Center, 2012).
- The owners of the St. Louis Cardinals baseball team sued Twitter when an unknown user created a fake Twitter account using manager Tony La Russa's name. The user posted comments, spiced with offensive language, prompting La Russa's complaint alleging that the fake Twitter page, "constituted trademark infringement and dilution, cybersquatting, and misappropriation of name and likeness"[1] (*La Russa v. Twitter, Inc.*, 2009).

"Murder most web 2.0" has been committed. Consider the conviction of Edward Richardson, who stabbed his wife to death in a "frenzied and brutal" attack in May 2008, precipitated by his wife's Facebook status adjustment from "married" to "single." Richardson was sentenced to 18 years in prison (British Man Killed Wife, 2009).

Hundreds of students have been disciplined for material they posted online. Mostly involving middle and high school students, discipline, expulsion from school, and lawsuits have resulted. The jurisdiction of school authorities is at the heart of these cases. Although some cases deal with students who write while on school premises, some post about school from "beyond the schoolhouse gate."[2] There is a growing body of case law beginning to carve out the rights and limits on discipline and students' free speech on social media.[3] There are no bright-line rules for determining when schools can and can't punish a student for statements made online.

Likewise, there is no bright-line rule for determining contributory copyright infringement. Pinterest, a popular photo sharing social network site in the United States, has attracted over 17 million users since 2009, allowing those users to curate boards of photos they post from their Internet searches and allowing other users to click on links to the original source. Other users may choose to repin the image on boards of their own. Pinterest has been under intense public scrutiny, with critics suggesting it is time to consider whether the site is facilitating copyright infringement on a mass scale, á la Napster a decade ago (Poletti, 2012).[4] Is there a legal difference between reposting, copying from a host server, and a user adding an image to his or her board that has already been posted to another user's board? Is there a legal or ethical concern associated with retweeting? This is called "repinning" and is somewhat analogous to "retweeting" a message on Twitter. Is reposting a copyright infringement? (Wassom, 2012).

Privacy concerns abound. Government oversight is beginning. Facebook reached a settlement with the Federal Trade Commission after being accused of "unfair and deceptive" practices. Facebook had shared user information with outside application developers, contrary to representations made to its users. And even after a Facebook user deleted an account, according to the Federal Trade Commission, the company still allowed access to photos and videos. The settlement reached in November 2011 requires the company to respect the privacy wishes of its users and subjects it to regular privacy audits for the next 20 years. Google's Street View project resulted in concern, litigation, and government investigations in Europe, Canada, and the United States when it was revealed that its vehicles had collected personal information from some Wi-Fi networks while photos were being taken for the project. In the United States, the Federal Communications Commission and the Federal Trade Commission both cleared Google of charges. However, Google was found to have violated Canadian law, and it remains under investigation in France and the Netherlands (Streitfeld & Wyatt, 2012).

Novel issues abound: Is "name squatting" (impersonating or using another individual's abandoned name) unethical? Who is responsible for content posted to a wiki? What is unfair or deceptive advertising when targeted marketing is possible? Do tweets, or statements on Facebook or other social media, qualify as an endorsement or testimonial in the realm of advertising regulation? Can reporters tweet from courtrooms?

The power of social media and the complexities of regulation were revealed in September 2012 when Google, owner of YouTube, selectively blocked the anti-Islamic video, "Innocence of Muslims" which mocked Islam and Prophet Mohammed. The film sparked protests and violence around the Arab world. The company quickly responded to violence in Libya and Egypt by voluntarily blocking access in those countries. In accordance with internal policies, Google blocked access in several countries including India, Indonesia, Singapore and Saudi Arabia after requests were made on behalf of those governments noting the film violated local law.

But Google Inc rejected a request by the White House to consider removing the YouTube clip thought to have ignited anti-American protests saying the video was within its guidelines. The case has led some to question whether today free expression is increasingly being determined by companies rather than governments (What to Make of Google's Decision, 2012).

## Service Providers Regulations: Terms of Service Agreement

The sources of regulation are diverse. Certainly governmental prohibitions and rules have been significant in many of the exemplars noted here, but "power of the traditional approach to regulation by nation-states may take a back seat to the influence of all-powerful 'terms of service agreements'" (Drucker et al., 2010, p. 68).

Terms of service and contractual obligation are the frontline in understanding the web of regulations in which social media operate. What percentage of social media/social networking users actually read the terms of use agreement or, for that matter, comprehended the terms? In haste, could you be disclosing sensitive or proprietary information? Every time we join a new social network, there is always a page or pop-up that requires us to read, acknowledge, and accept the network's Terms of Service agreement. Unfortunately, most of us never take the time to read this agreement. As a result, we agree to terms with which we may not be comfortable. Key elements of Terms of Service agreements that the user should understand include the outline the service provides, details of the type of content the user can share or download, a statement regarding who owns content provided by the user, and rules of required conduct. Terms of Service agreements typically remind users they are not exempt from the applicable laws and regulations which, in an online environment, may cause confusion with regard to which government's laws apply.

Policies and practices are significantly different among social networking sites. Twitter "claims no intellectual property rights over the material [users] provide" and further states that users "can remove [their] profile at any time by deleting [their] account" and that doing so "will also remove any text and images [users have] stored in the system" (Twitter, 2012). By contrast, LinkedIn requires its users to grant LinkedIn a "nonexclusive, irrevocable, worldwide, perpetual, unlimited, assignable, sublicenseable, fully paid up and royalty free right…to copy, prepare derivative works of, improve, distribute, publish, remove, retain, add, and use and commercialize, in any way now known or in the future discovered…without any further consent, notice and/or compensation to you or to any third parties" with respect to all submitted content (LinkedIn, 2012). Most social networking sites address the use of third party content in their terms of use. Twitter's "Basic Terms" include the statement that users are "solely responsible for…any data, text, information, screen names, graphics, photos, profiles, audit and video clips, links that the user submits, posts or displays" (Twitter, 2012). Another basic term states that users "must not, in the use of Twitter, violate any laws in your jurisdiction (including but not limited to copyright laws)"(Twitter, 2012). Therefore, in addition to violating copyright laws, the unauthorized use of third-party content violates the terms of use of most social networking sites and could serve as an additional basis for liability. Social networks generally prohibit the posting of defamatory content by members and reserve the right (while not assuming the responsibility) to remove any such content (Twitter, 2012).

## Governments: Blocking and Prohibitions

Social networks challenge organizational, educational, and military[5] contexts. The leaders of authoritarian countries have targeted social media sites seen as threats to existing political power. Blocking access to "undesirable" sites has been a common government tactic. Many of the blocked or blacklisted sites contain "subversive" content on topics ranging from religion and sex to women's rights and popular culture. Social media sites frequently blocked include eBay, Facebook, Bebo, Myspace, orkut, YouTube, Pandora, Photobucket, Yahoo! Messenger, AOL AIM, Flickr, and many blogs.

Social media are particularly significant sites of political ferment in countries in which traditional mass media are controlled by the government. Citizen journalists, user-generated content, and conversation thrive, but increasingly, governments have responded by attempting to exert greater monitoring, control of access, and punishments. According to Reporters Without Borders, more than 120 cyberdissidents are being detained for expressing their views online as of the spring of 2012. In the report, *Enemies of the Internet* (2012), it is noted that while the Arab Spring highlighted the potency of social networks as tools for protest, the result was months of increased repression and "tougher measures to what they regarded as unacceptable attempts to 'destabilize' their authority" (*The Enemies of the Internet*, 2012). Politically motivated blockages abound. For example, at the end of February 2011, Egypt cut Internet access for five days.

Slowing the Internet connection speed right down is more subtle but also effective as it makes it impossible to send or receive photos or videos. Iran is past master at this. Syria's censors also play with the Internet connection speed, fluctuations being a good indicator of the level of repression in a given region (*The Enemies of the Internet*, 2012).

In Belarus, access was blocked from the social network Vkontakte during a period of demonstrations in that country. Tajikistan has blocked Facebook.

As soon as the uprisings in Tunisia and Egypt got under way, most regimes that censor the Internet quickly reinforced online-content filtering in a bid to head off any possibility of similar unrest spreading to their own countries. Some regimes have adopted filtering as a standard tool of governance, one that strengthens their hold on power. Livestreaming sites and social networks are often the most affected (*Enemies of the Internet*, 2012).

While social media did not create the Arab Spring, the countries involved (e.g.,Tunisia to Egypt to Libya) did have a growing civil society movement facilitated or complemented by online activism.

Just after the streets of Tunisia and Egypt erupted, China saw a series of "Jasmine" protests, prompting the government to crack down further on Internet freedoms. China is becoming the largest Internet market in the world, but Facebook and Twitter are blocked in China. Despite this fact, millions of Chinese users access the social network by using anonymous IP servers.[6]

Much has been written about the so-called Great Firewall of China, which precludes Chinese residents from accessing foreign Web sites such as Google and Facebook. But increasingly, it is impossible to block access to all sites from within or outside a country and maintain useful function of the global reach of the Internet for commercial purposes. "An alternative is to allow access to sites, but police the content, eliminating messages deemed objectionable. Automated methods may be used to eliminate some messages, while others are deleted manually" (China's Social Networks, 2012). A study of "soft censorship" of social media in China, conducted by Noah Smith, David Bamman, and Brendan O'Connor, appears in the March issue of *First Monday*, which is focused on Chinese social media sites, particularly microblogs or weibos.[7] An analysis of almost 60 million messages from China's equivalent of Twitter indicated that certain topics were banned and not all of these were politically sensitive in nature. The system "stops people visiting some sites outside China, returns no results for searches of banned terms, censors chat and vets blogs. Banned topics include the Falun Gong spiritual movement and human rights activist Ai Weiwei" (China's Social Networks, 2012).

## Companies: Cooperation and Compliance

Technology companies have also been complicit in working within national censorship laws. In 2012, Twitter and Google announced that they would selectively censor information by country. Headlines proclaimed, "Twitter Announces Micro-Censorship Policy" (Sengupta, 2012). The company explained that previously removing a tweet would result in worldwide removal, but a newly developed capability to selectively remove content country by country allows Twitter to comply with requests within a country to observe local law but not remove the tweet elsewhere. Twitter noted it will attempt to let users know if their tweet is being blocked within their country and will clearly mark when the content has been withheld. Google followed shortly after. Bloggers who relied on Google to organize protests during the Arab Spring found their posts being blocked by Google itself. The company will now block posts or blogs from being seen in a particular country if they violate local laws. Internet freedom group Open Net Initiative said, "the change marks a new trend in American Internet companies bowing to the demands of authoritarian regimes"(Waugh, 2012). In India, authorities are bringing pressure on service providers, trying to persuade them to provide a preview of content so that anything "shocking" or liable to provoke sectarian strife can be deleted.

## Ethical Questions

Corporate compliance with authoritarian regimes raises ethical issues. Should students and teachers ever be friends on Facebook? (Should Teachers, 2012)? Should judges permit attorneys to check the blogs and Web sites of prospective jurors? Should reporters friend sources? Should we manipulate our online identity? Should one ever use a pseudonym online? Is it ever ethical to post negatively about others at work? When should you accept friend requests on Facebook, LinkedIn, or Google+? Should you follow clients on Twitter, blogs, or Flickr? When is it appropriate to repost or pass information along? These questions merely scratch the surface. The scope of this volume is broad and, if successful, hopes to provide the reader with paths to follow through the social media labyrinth. To some extent, a large question needs to be grappled with— Is it possible for the informed citizen to function outside the parameters of the social media world? Depending on how one defines social media, the answer is a probable "no." Does the individual have the right to function outside the social media realm? The answer is a probable "yes," but it represents a highly unlikely possibility. Contemporary life imposes a technological presence from which there are few escape options.

# A Preview of the Volume

The growth of social media among all segments of the population in a relatively brief period of time has been astounding. "Friending," "Following," "Liking," and messaging shape our personal and professional lives. By definition, the scope of this volume is limited. It seeks to provide an introduction to key legal and political issues that have developed to date. The chapters that follow emphasize developments in U.S. contexts while noting the global nature of the subject at hand.

In Chapter 2, Star Muir considers the people using social media and what makes digital natives distinct. He explores the use of social media for political activism post Arab Spring and considers diverse attempts to regulate, particularly through terms of service and provider policies. As with any new media development, there are positive and negative consequences, intended and unintended uses.

In "Swimming in Cybercesspools: Defamation Law in the Age of Social Media" (Chapter 3), Dale Herbeck examines fundamental questions of protection from government regulation and potential liability for service providers within the context of emerging defamation law. He asks the all-important question: "Is speech on the new social media entitled to the same First Amendment protection as speech on the traditional media?"

In Chapter 4, we turn to the developing phenomenon of cyberharassment and cyberbullying. Juliet Dee provides a series of vivid cases of this type of psychological and sometimes physical assault. She provides an overview of case law and legislative action taken to address this new form of abuse. At what point do cyberbullying and harassment online and offline converge and diverge?

In Chapter 5, Mary Ann and Eric Allison further extend this examination of cyberbullying and discuss the unintended use of social media—sexting. Emphasizing ethical concerns, this chapter features the history of unsafe adolescent communication and suggests that legal responses should consider research on brain development relevant to processing consequences of behavior.

In Chapter 6, Adrienne Hacker-Daniels explores another unintended consequence of user-generated content in her examination of the implications of WikiLeaks. Focusing on the repercussions, reactions, and vague legal standing, she raises significant questions about user-generated content, service-provider responsibility for the content of others, and the new environment of a data dump. While WikiLeaks has embarrassed and threatened governments around the world, this chapter focuses on the data dump associated with the "Iraq War Diaries" and reaction within the United States.

One of the most contentious and examined areas of law and ethics considered to date has been the matter of privacy. Certainly, privacy implications of social media emerge throughout the volume. Warren Sandmann (Chapter 7) raises a quintessential question in asking, "What is the role of the private in the age of the social?" He approaches the pivotal question by going back to the 1890 essay written by Louis Brandeis and Samuel Warren for the *Harvard Law Review*. Thus the question raised in a preelectronic era remains an important and vastly more difficult one to answer. While the right to privacy remains a critical question that must constantly be posed, it has been reduced to a philosophical query because it is no longer obtainable. But the question must remain in the foreground.

Closely linked to the matter of privacy is the issue of transparency raised by Douglas Strahler and Thomas Flynn (Chapter 8). This chapter examines privacy from a different perspective and considers how social media creates a more transparent society with both positive and negative consequences for individuals, companies, and governments. The authors consider information availability and the transformation of the expectation of privacy.

In Chapter 9, Bruce Drushel examines the regulation of sexually explicit content. He surveys the relevant historical developments in which zoning or channeling communication has been a preferred approach and considers application of the well-tested mechanism in the social media context. Chapters 10 and 11 feature the novel ethical issues faced by media professionals confronted with new social media options. Suzanne Berman provides an overview of some of the public relations strategies used in a social media environment and outlines the ethical considerations that arise as a result of them. Definitions of social media public relations practices are provided as well as the ethical concerns they raise. Similarly, Kelly Fincham explores the journalistic uses of social media and the concomitant ethical challenges raised. Can old rules of ethics serve the new social media? How are media organizations approaching social media policies? How can journalistic credibility be maintained while taking full advantage of social media?

The volume concludes with a visit to our crystal ball and the consideration of future issues in law and ethics as social media environments mature.

Howard Cohen, communication technologist, has warned

There is so much being written about social media and social networks that it is really impossible to produce a "point-in-time" report on the "state" of social media. It is truly a moving target, and a very quickly moving target at that." (Drucker, Gumpert & Cohen, 2010, p. 68).

This volume is an attempt to document efforts to bring the law and ethical standards up to speed with technological and social realities. In that spirit, the following pages offer a snapshot in time of the state of the legal and ethical landscape of social media.

## Notes

1. Shortly after the lawsuit was filed, Twitter took down the fake profile. The case was settled in June 2009.
2. The U.S. Supreme Court has made clear that physical boundaries alone are not the determining factor in the jurisdiction of student behavior. The 2007 case of *Morse v. Frederick*, the so-called BONG HITS 4 JESUS case, allowed the school to remove the sign even though the student was standing across the street from school property.
3. Several U.S. court decisions concerning disciplinary action by public schools have actually reversed the school administrative decisions on the grounds that the students' online posts were speech protected by the First Amendment. *Layshock v. Hermitage School District* (2010) and *J.S. v. Blue Mountain School District* (2010) were decided by the U.S. Court of Appeals for the Third Circuit in June 2011. In *Layshock*, the court overturned the suspension of a high school student who, at home, created fake profiles for his principal on Myspace, containing "unflattering" content. Absent a showing of disruption at school, the court "reject[ed] out of hand any suggestion that schools can police students' out-of-school speech by patrolling the public discourse." Likewise, in *J.S.*, the student plaintiff, a middle school student, suspended for creating a fake Myspace profile for her principal, containing sexually explicit content, had not caused an in-school disruption. In January 2012, the U.S. Supreme Court decided not to weigh in on the free speech rights of students on the Internet.

    If content originates from a classroom and or directly relates to school activities, it is easier for schools to assert jurisdiction (*Doninger v. Niehoff*, 2009).
4. Pinterest has a procedure in place to try to deter copyright infringement by enabling content owners to report a violation and have the content taken down. Some argue pinning copyrighted photos easily found on the Internet constitutes "fair use" if the images' creators are credited or if a link goes back to the site of origin. Other photo sharing sites like Yahoo's Flickr have added an opt-out code for users, who can disable images and bar them from being shared on Pinterest (Poletti, 2012).
5. The U.S. military lifted its ban on social media in January 2011. Commanders at any level can still institute a temporary ban on specific sites. Along with the ban lift came specific branch policies for social media use (*U.S. Army Social Media Handbook*, 2011).

6. This is the situation in other countries, including Iran, where Facebook is banned by the government.
7. The Carnegie Mellon team analyzed almost 57 million messages posted on Sina Weibo, a domestic Chinese microblog site similar to Twitter that has more than 200 million users (Carnegie Mellon Performs, 2012).

## References

Agarwal, S. (2011, Dec. 21). Mobile social media: The next great disruptive trend. Retrieved from http://technorati.com/social-media/article/mobile-social-media-the-next-great/
Alleyne, R. (2011, April 19). The young generation are "addicted" to mobile phones. Retrieved from http://www.telegraph.co.uk/technology/8458786/The-young-generation-are-addicted-to-mobile-phones.html
boyd, d. m., & Ellison, N. B. (2007). Social network sites: Definition, history, and scholarship. *Journal of Computer-Mediated Communication, 13*(1), X. Retrieved from http://jcmc.indiana.edu/vol13/issue1/boyd.ellison.html
British man killed wife over "single" Facebook status. (2009, Jan. 23). Retrieved from http://www.huffingtonpost.com/2009/01/23/british-man-killed-wife-o_n_160453.html
Bullas, J. (2012). 20 stunning social media statistics plus infographic. Retrieved from http://www.jeffbullas.com/2011/09/02/20-stunning-social-media-statistics/
Carnegie Mellon performs first large-scale analysis of "soft" censorship of social media in China (2012, March 7). Retrieved from http://www.cmu.edu/news/stories/archives/2012/march/march7_censorshipinchina.html
Castells, M. (1989). *The informational city: Information technology, economic restructuring, and the urban-regional process*. Oxford, UK: Blackwell.
China's social networks hit by censorship, says study. (2012, March 9). Retrieved from http://www.bbc.co.uk/news/technology-17313793
Cohen, H.M., (2012, March 26), personal communication.
comScore (2011, Nov. 21) Mobile social networking audience grew 44 percent over past year in EU5. Retrieved from http://www.comscore.com/Press_Events/Press_Releases/2011/11/Mobile_Social_Networking_Audience_Grew_44_Percent_Over_Past_Year_in_EU5?utm_source=feedburner&utm_medium=feed&utm_campaign=Feed%3A+comscore+%28comScore+News%29&utm_content=Google+Reader
Concern about health of woman journalist held for Facebook comments.(2012, April 6). Retrieved fromhttp://en.rsf.org/palestinian-terr-concern-about-health-of-woman-06-04-2012,42271.html
Dictionary.com. (2012). Retrieved from http://dictionary.reference.com/browse/social+media?s=t
*Doninger v. Niehoff*, 594 F. Supp. 2d 211 (2009).
Drucker, S., Gumpert, G., & Cohen, H. (2010). Social media. In S. Drucker & G. Gumpert (Eds.), *Regulating convergence* (pp. 67–102). New York: Peter Lang.
Fired over Facebook: Employer regulation of speech on social media sites (2011, Feb. 25). International Bar Association, Retrieved from http://ibamedialaw.wordpress.com/2011/02/25/fired-over-facebook-employer-regulation-of-speech-on-social-media-sites/
Fishman, D. (2010). Media convergence and copyright issues." in S. Drucker & G. Gumpert (Eds.), *Regulating convergence* (pp. 123–146). New York: Peter Lang.

Gottlieb, J. (2011, March 11). Ex-girlfriend's phony Facebook page tests NJ identity theft law. Retrieved from http://socialmedialawnews.com/

*Houchins v. KQED*, 438 U.S. 1 (1978).

Jenkins, H. J. (2006). *Convergence culture: Where old and new media collide*. New York: New York University.

*J.S. v. Blue Mountain School District*, 593 F.3d 286 (3d Cir. 2010).

*La Russa v. Twitter, Inc. Citizen Media Law Project*. (2009). Retrieved from: http://www.citmedialaw.org/threats/la-russa-v-twitter-inc

*Layshock v. Hermitage School Dist.*, 593 F.3d 249 (3d Cir. 2010).

LinkedIn. (2012). Retrieved from http://www.linkedin.com/static?key=user_agreement&trk=hb_ft_userag

McNaughton, M. (2012, March 29). 77% of German internet users access social media via mobile. Retrieved from http://therealtimereport.com/2012/03/29/77-of-german-internet-users-access-social-media-via-mobile/

Merriam-Webster Dictionary. (2012). Retrieved from http://www.merriam-webster.com/dictionary/social%20media

*Morse v. Frederick*, 551 U.S. 393 (2007).

New Knight Center Twitter feed showcases restrictions on social media freedom. (2012). Retrieved from http://knightcenter.utexas.edu:8080/blog/new-knight-center-twitter-feed-showcases-social-media-freedom

Poletti, T. (2012, March 14). Is Pinterest the next Napster? *The Wall Street Journal*. Retrieved from http://online.wsj.com/article/SB10001424052702304450004577279632967289676.html

Reporters Without Borders (2012, June 2). Retrieved from: http://en.rsf.org/

Sengupta, S. (2012, Jan. 26). Twitter announces micro-censorship policy. *New York Times*. Retrieved from http://bits.blogs.nytimes.com/2012/01/26/twitter-announces-micro-censorship-policy/

Should teachers and students be Facebook friends? (2012, April 19). Fox News. Retrieved fromhttp://www.education.com/magazine/article/Students_Teachers_Social_Networking/

Solis, B.(2011). Who is the me in social media? Retrieved from http://www.briansolis.com/2010/01/who-is-the-me-in-social-media/

Streitfeld, D., & Wyatt, E. (2012, April 16). Unanswered questions in F.C.C.'s Google case. *New York Times*, p. B1.

The Enemies of the internet. (2012). Reporters Without Borders. Retrieved from: http://march12.rsf.org/en/

The history of postcards. Retrieved from http://www.emotionscards.com/museum/historyofpostcards.htm

The World UNPLUGGED project. (2012). Retrieved from: http://theworldunplugged.wordpress.com/about/

Twitter. (2012). Retrieved from http://twitter.com/tos

U.S. Army social media handbook. (2011). Retrieved from http://www.google.com/search?client=safari&rls=en&q=US+Army+SOcial+Media+Handbook&ie=UTF-8&oe=UTF-8

Van Grove, J. (2011, Oct. 20). Social networking on mobile devices skyrockets. Retrieved from http://mashable.com/2011/10/20/mobile-social-media-stats/

Wassom, B. (2012, April 12). Pinterest and copyright: Repinning is not reproduction. Retrieved from http://www.wassom.com/pinterest-and-copyright-repinning-is-not-reproduction.html

Waugh, R. (2012, Feb. 3). Google joins Twitter in censorship storm: Site may now block blog posts in line with requests from oppressive regimes. Mail Online. Retrieved from http://www.dailymail.co.uk/sciencetech/article-2095328/Google-joins-Twitter-censorship-storm-Site-block-blog-posts-line-requests-repressive-governments.html

What to make of Google's decision to block the 'Innocence of Muslims' movie (2012, Sept. 14). *The Atlantic,* ttp://www.theatlantic.com/technology/archive/2012/09/what-to-make-of-googles-decision-to-block-the-innocence-of-muslims-movie/262395/

# · 2 ·

# Privacy, Identity, and Public Engagement among Digital Natives

Living with teenage digital natives is at once a wonderful and daunting experience, fraught with the thrill and drama of connection, the challenge and stimulation of endless amusements, the reality of growing up with a portal to the largest information bank in the known universe, and yet rife with implications, with unintended and intended consequences. Being cast as a digital immigrant, a visitor from a time of card catalogs, typewriters, and payphones, educated in an era of "tell and test," puts a particular slant on my interest in digital natives and their evolving social environments. As an immigrant, I feel a need to understand a bit more about how my children and the bulk of their generation are both empowered and disempowered by their new tools.

## I. Digital Natives *in situ*

When Marc Prensky (2001) coined the term *digital native*, he likely didn't have any idea how it would catch on. A Google search now reveals 5.9 million hits for the phrase, along with 64 books listed on Amazon.com, and a raging academic controversy. An important part of his initial framing is the distinction between natives and immigrants, between people grown up with the internet and people who remember and were shaped by a time before widespread mediated technologies.

## A Generation of Digital Natives

There have been quite a few larger projects focused on characterizing this new Net Generation. Tapscott's team surveyed over 10,000 youth in 12 countries, and identified eight Net Generation norms, among them freedom (to change jobs or work outside the office), collaboration (enjoying community and influencing decisions), entertainment (often multiple concurrent "netivities"), and speed (just-in-time, 24/7) (2009, pp. 73–96). Palfrey and Gasser (2008) similarly document a portrait of qualities, and while expressing concern about multitasking (with a loss of focus), short attention spans, and a "copy and paste" culture, they try to address immigrant fears about safety and privacy and prevent overreactions, which result in weak and inflexible solutions.

## Digital Natives in School

Prensky's original distinction between natives and immigrants was in the context of the disconnect between the media-rich environment of the digital native and the immigrant-created educational system featuring talking heads and linear delivery. Jukes, McCain, and Crockett (2010) follow Prensky's (2010) lead in questioning current approaches to K-12 education to avoid an impending "disconnect tragedy." They trace evolving functional differences in brain structure, reading patterns, and learning styles to explain why students are bored in school and have very different learning preferences (pp. 35–41). Prensky's partnering pedagogy is similar to problem-, inquiry-, and case-based learning and underscores the need to tap into student passions to drive learning that is not only relevant but real in the student's world.

Many immigrants continue to have reservations about digital natives, some based on experience, some based on research.

## The Dumbest or the Most Distracted Generation?

Bauerlein (2008) offers a Bloom-esque lament about the state of the younger generation: a near total disconnect to their civic and cultural inheritance. Peer opinion matters more than adult tutelage, and the immediate triumphs over history. Test scores are going down, no one knows their basic geography anymore, few possess knowledge or inclination to be an informed citizen, and an anti-intellectual outlook prevails in their leisure lives. Tapscott (2009) counters Bauerlein with evidence of stronger digital native visual acuity, more capable multitasking, better intensive problem-solving, and far better information processing: "[G]rowing up digital has equipped these Net Geners with the mental skills, such as scanning and quick mental switching, that they'll need to deal with today's overflow of information" (p. 118).

There are certainly skills and habits of mind that immersion in digital environments has developed and created. An emerging literature refers to them as "the most distracted generation." Oppenheimer (2003) offers this summary of a generation with "flickering minds":

> America's students…have become a distracted lot. Their attention span—one of the most important intellectual capacities anyone can possess—shows numerous signs of diminishing. Their ability to reason, to listen, to feel empathy, among other things, is quite literally flickering. (p. xx)

Jackson (2009) examines the effects of a culture of digital distraction on students, and she identifies impacts on critical thinking, problem-solving skills, deep relationships, and cultural memory. Her work on focus and attention leads her to predict a coming dark age, since we are "eroding our capacity for deep, sustained, perceptive attention—the building block of intimacy, wisdom, and cultural progress" (p. 13). Because "our virtual, split-screen and nomadic era is eroding opportunities for deep focus, awareness, and reflection…we face a real risk of societal decline" (p. 25). Tracing insights from Plato to McLuhan, and reflecting on his own media scanning habits, Carr (2010) argues in *The Shallows: What the Internet Is Doing to Our Brains*, that every technology carries an "intellectual ethic," and the internet leads to rapid sampling of information bits from many sources. The brain begins to receive positive reinforcement for each new click and scan, which over the long run, by accustomation, makes reading extended articles or books more difficult.

Focusing more specifically on youth and young adults, several works explore the implications of a changing media environment for our evolving brains. Rosen, Carrier, and Cheever (2010), in *Rewired: Understanding the iGeneration and the Way They Learn*, report that the iGeneration and Net Generation are far more likely to multitask than Baby Boomers and Gen Xers, and they handle up to six tasks simultaneously with greater ease, all of which function to keep them interested. All still suffer from levels of dual task interference, but the authors suggest that educators meet the need for speed and multitasking by using mobile technologies, engaging interest and passion via social networking, and considering video games to motivate learning.

Small and Vorgan (2008), authors of *iBrain: Surviving the Technological Alteration of the Modern Mind*, argue that the human brain is evolving faster now than at any previous time in history. Small directs a neuroscience center on memory and aging at UCLA, and the authors have investigated, among other issues, impacts of multitasking and multiple channels of stimulation, including continuous partial attention (which can be effective for a while but over time causes "digital fog"—fatigue, irritability, distraction). Tracing the rise of digital "stimulus junkies" through constant exposure to the internet and multiple information streams, youth develop habits of mind that undermine reflective evaluation:

> They no longer have time to reflect, contemplate, or make thoughtful decisions. Instead, they exist in a sense of constant crisis—on alert for a new contact or bit of exciting news or information at any moment. Once people get used to this state, they tend to thrive on the perpetual connectivity. It feeds their egos and sense of self-worth, and it becomes irresistible. (Small & Vorgan, 2008, p. 18)

These characterizations of digital natives have not gone unchallenged. Even as there are serious concerns expressed about the mental habits, attention spans, and cultural memories of youth growing up with the internet, there are also direct critiques of the implicit determinism and associated pedagogical practices with these characterizations.

## Hasty and Ill-Advised Generalizing?

A major challenge to the digital native characterization has come from scholars in Britain, Australia, and South Africa. They decry stark generalizations about youth as more tech savvy and note that there are often larger differences in technology use within this group than between this group and other generations (Burhanna, Seeholzer, & Salem, 2009; Hargittai, 2010; Margaryan, Littlejohn, & Vojt, 2011). Based on survey data, Kennedy, Judd, Dalgarnot, and Waycott (2010) conclude that Net Generation students are far from homogenous, that use of one technology cannot reliably predict use of others, and that other demographic variables (gender, university, and cultural background) are better predictors of a student's technology experience (p. 341). In South Africa, Brown and Czerniewicz (2010a) ultimately conclude that this characterization of a population of students is in fact descriptive of an elite group. Lamenting the emphasis on educational practices focused on this elite, they term it *digital apartheid* and argue that educational resources, as a matter of justice, should be focused on uplifting the "digital strangers" who are lagging initially in skills and are falling farther behind as they prioritize their communication technology use (2010b, p. 364).

Palfrey and Gasser (2008) readily concede that they tend to think of this group less as a *generation* and more as a *population*. The latest iteration of the Educause Center for Applied Research (ECAR) annual report, a longitudinal survey of over 36,000 young adult college students, indicates that from 2008 to 2010, daily text messaging increased from 50% to 75%, and that Facebook use increased from 89% to 97% (Smith & Caruso, 2010, p. 55). More than 80% own a laptop computer, and nearly all (99%) own at least one computer (Smith & Caruso, 2010, p. 37).

There seems to be some truth that both access to and extensive use of digital technologies vary across geography and other variables and that there are dominant use patterns that make it valuable to refract issues through a set of population characteristics.

## Facebook as the Stage of Life

In 2010, Facebook hit two significant milestones: it registered over half a billion users (and over $1.3 billion in revenue; Baig & Swartz, 2010), and it became the most visited web site on the planet, out-performing Google and Yahoo and garnering 8.9% of total internet traffic (Mui & Whoriskey, 2010).[1] The latest self-reported Facebook statistics show that the average user has 130 friends, and that people spend over 700 billion minutes per month on Facebook. There are now 900 million objects for interaction and more than 30 billion pieces of content available on Facebook. The average user is connected to 80 community pages, groups, and events, and creates 90 pieces of content each month ("Press Room," 2011).

Teen and young adult use has been a powerful driving force behind many of these 700 billion monthly minutes. Lenhart, Purcell, Smith, and Zickuhr (2010) of the Pew Internet and American Life Project report that for the latest data (2009), "93% of teens ages 12–17 go online, as do 90% of young adults ages 18–29" (p. 2), and "nearly three quarters (73%) of online teens and an equal number (72%) of young adults use social network sites" (p. 4). Teens and young adults are consistently the highest user groups for the internet and for social networking. As part of their 7.5 hours a day of media use (11 total hours if you count concurrent use; Mundy, 2010), 63% go online every day (including 36% several times a day; Lenhart, et. al., 2009, p. 7). Valenzuela, Park, and Kee (2009) report that by college, up to 95% of students may have a Facebook account (p. 894).

Teen use of Facebook runs a wide gamut of applications, preferences, and behaviors. The Nielsen report titled, "How Teens Use Media" (Nielsen Company, 2009), identifies a close similarity between adult and teen use patterns, in terms of categories explored and search needs, but it notes one important difference, that 57% of teens said they consult their social networks for advice, which is 63% more likely than the average Facebook user (Nielsen Company, 2009, p. 7).

There are several literatures involved in the discussion of how Facebook functions for users, including uses and gratifications research. In a recent comparison of Facebook and instant messaging, Quan-Haase and Young (2010) identify six key motivations for using Facebook (pastime, affection, fashion, the sharing of problems, sociability, and the gathering of social information) and concluded that Facebook is more about having fun and knowing about the social activities occurring in one's social network, whereas instant messaging is geared more toward relationship maintenance and development. Park, Kee, and Valenzuela (2009) identify four similar dimensions: socializing, entertainment, self-status seeking, and information. In a comparison between the first three and information use, they conclude that most Facebook group use is for socializing, entertainment, and status, yet it is information use that strongly predicts offline political participation. Integrating groups better into community issues, they conclude, will enhance political empowerment.

While uses and gratification studies can offer a dizzying array of categories and user motivations, Bonds-Raacke and Raacke (2010) identify an information dimension, a friendship dimension, and a connection dimension, and they also note a clear gender effect: males use the social network site significantly more for dating, and females set their controls to private significantly more often.

Another growing area of research is on the use of social networks for identity formation. A personal profile, boyd (2011) observes, is a place where people write their self into being (p. 43). Palfrey and Gasser (2008) explain two paradoxes of teen identity formation on the internet: First, digital natives can adjust their identities with ease, but they have less control over how others perceive their identity than in other eras; and second, although digital natives can create multiple identities with little effort, they are more bound to a single identity than before. The affordances of the new technologies, boyd (2011) argues, contribute to these paradoxes. Moving from an analog world with atoms into a digital world with bits, she notes, creates several affordances that reflect the nature of bits: content is *persistent* (recorded and archived), *replicable* (can be duplicated), *scalable* (visible to larger numbers of people), and *searchable* (accessed systematically; p. 46). Hence the ease of profile adjustment also involves scalability and a loss of control over how others might view your identity and particularly an identity that is persistent over time. Grasmuck, Martin, and Zhao (2009) explain that social networking offers a chance to create a "hoped for" self (p. 158), while Ball (2010) terms Facebook the "mirror on the wall" (p. 2).

This leads us toward a perspective on Facebook for teens as increasingly creating their "stage" of life. Gozzi (2010) writes that the "stage for the drama [of life] is no longer a television show, as I expected in the 1990s. The stage is Facebook" (p. 235). Page (2010) offers an extended analysis of narrative form in status updates among Facebook friends. The status archive, she argues, upends the overarching structure associated with canonical narratives and replaces it with episodic narrative, only loosely connected by chronology:

> The networking capacity of digital technology enables the multi-dimensional distribution of status updates into numerous, constantly changing sequences of Facebook activity, generating a networked web of story episodes rather than a single series of events with a defined point of inception and closure. (p. 441)

This seems consonant with digital native characteristics previously identified by Small and Vorgan, Rosen et al., Jackson, and Carr. Within an episodic framework, greater value is attached to recency over retrospection. The self-portrait that emerges from the archive, Page analogizes, is an identity that is "continuously being created using a pointillist technique," which deviates from linear connections that give a coherent whole and invites strategies that "fill in the gaps" of an evolving account of the drama of their lives (p. 440). It is this "dramatization" of life applied to the digital environment that brings us to deep and continually emerging immigrant concerns about risks young digital natives face on this stage.

## II. Immigrant Fears and the Regulatory Environment

Termed the equivalent of a "new Wild West" (Lee, 2008, p. E1), there is an ongoing relationship between fears about internet usage (and particularly social networking) and calls for local, state, and federal action to combat these concerns. This section discusses the progressive blurring of the distinctions between public and private information and strongly related parental (and societal) fears about the abuse of information power (including cyberbullying, sexual predation, and addiction), which point toward a summary of the regulatory environment emerging to address social networking dysfunctions.

## The Blurring of Public and Private Spheres

Of particular concern to parental immigrants is that many teenagers, in the context of a circle of friends, now publicly reveal information that their parents would never have considered making public. Early work done by Gross and Acquisti (2005) indicates that 91% of profiles contain an image, 88% of users reveal their birth date, 40% list a phone number (29% list a cell phone number), and 51% list their current residence These disclosures are verified in international settings through extended profile analysis (Taraszow, Aristodemou, Shitta, Laouris, & Arsoy, 2010), but more in-depth interviews reveal that teens do care about privacy, only from a different perspective than their parents:

> Teenagers described thoughtful decisions about what, how and to whom they reveal personal information....This suggests a definition of privacy not tied to the disclosure of certain types of information, rather a definition centred on having control over who knows what about you....(Livingstone, 2008, p. 404)

Palfrey (2010) agrees that teens are concerned about privacy in some very specific ways, including keeping certain information from their parents and teachers. The main issue here, as Gelman (2009) points out, is that Facebook and other social network sites are "blurry-edged," and while the technology provides an "illusion of privacy and control," this becomes problematic in at least two ways (p. 1319).

First, the dramatic nature of these sites ensures that the stories that are told involve more than one person. That is because, as Gelman (2009) observes, "one engaging aspect of people's stories is that they usually involve others" (p. 1318), and although "stories are much richer when we identify other participants in our lives," this richness "comes with a price" (p. 1328). There are incentives built into the affordances of Facebook as a community-builder, Gelman concludes, that push each member to disclose more. The result, Bahadur (2009) explains, is a "digital combination of a billboard and a scrapbook; a resume and a diary; a tabloid magazine and a family photo album; and a reality television show and a family video all rolled into one" (p. 344). The underlying difficulty is that while the tabloid and the diary may be intended for a circle of friends, neither the social nor the legal framework can prevent "friends" from discussing, linking, or posting material (particularly spice or drama). The privacy violations, Grimmelmann (2009) concludes, "emerge spontaneously from the natural interactions of users with different tastes, goals, and expectations. The dark side of a peer-to-peer individual-empowering ecology is that it empowers individuals to spread information about each other" (pp. 1188–1189).

A second related issue is the misperception that people have regarding privacy controls. In-depth interviews allowed Debatin, Lovejoy, Horn, and Hughes (2009) to conclude that while most Facebook users claimed to understand privacy issues, they uploaded vast amounts of personal information and failed to clearly understand specifics about Facebook privacy controls. For most, the risks of privacy violations were ascribed to others (rather than the self), and the benefits of online social networking outweighed the risks of disclosure (p. 100). The risks of online social activity are very real, however, and while the media may fuel these fears with lurid headlines, parents are legitimately concerned about the blurring of private and public spheres.

## Abuse of Information Power

In this wild digital frontier, one of the most frequently identified threats is online bullying, with several recent suicides notable and tragic outcomes of extended campaigns of abuse and a recent college student suicide prompted by online posting of a video of homosexual activity (see Hoover, 2006; Jones, 2008; Khadaroo, 2010; Lappin & Eglash, 2011; Lee, 2008). Online bullying is often particularly pernicious because of the blending of private and public spheres and the public nature of the embarrassment that previously had been limited by line of sight.

The speed and reach of the internet, Meredith (2010) explains, make cyberbullying more rapid and potentially more public. In their study of data from 7,300 adolescents, Wang, Nansel, and Iannotti (2011) report that although cyberbullying is the least prevalent form of bullying—physical (21.2%), verbal (53.7%), and relational (51.6%) bullying were more common than cyberbullying (13.8%; p. 416)—they also found that for cyberbullying, there was noticeably less depression for bullies and bully victims and substantially higher rates of depression for victims of cyberbullying. This poses obvious challenges, as Hoffman (2010) writes: "The lawlessness of the Internet, its

potential for casual, breathtaking cruelty, and its capacity to cloak a bully's identity all present slippery new challenges to this transitional generation of analog parents" (p. A1).

Perhaps the worst parental fears are not about bullying but about vulnerability to sexual stalkers and predators. Police and child safety experts warn that
"sexual predators, pornographers and prostitution rings are capitalizing on the
rising popularity of mobile devices and social media to victimize children"
(Acohido, 2011, p. 1A). This is clearly an area where the hype exceeds expectations of risk. Wolak, Finkelhor, Mitchell, & Ybarra (2008) document that publicity about sexual predators online is largely inaccurate and that most of the
situations that arise fall under statutory rape rather than child molestation or
forcible sexual assault (see also, Mitchell, Finkelhor, Jones, & Wolak, 2010.) Yet
the seriousness of the issue led Facebook to work with state law enforcement officials to voluntarily ban sex offenders from membership and to make it more difficult for members to search for underage members (Worden, 2008).

Another parental fear concerns the disproportionate amount of time
teenagers spend online in their social networks. Internet addiction has become
a source of controversy in the mental health field. Some psychiatrists, like Dr.
Jerald Block, argue that the disorder is so common that it should be included
in the bible of mental problems, the *Diagnostic and Statistical Manual of Mental
Disorders* (Smith, 2008). In South Korea and Malaysia, the problem has been
diagnosed on a large scale. Dr Mohamad Hussain Habil, director of Universiti
Malaya Centre of Addictive Sciences, believes there is a problem with imbal-
anced Facebook usage, or "Facebook Addiction Disorder": "It is similar to
Internet addiction. It is a type of behavioural addiction similar to pathologi-
cal gambling, sexual and shopping addiction" (Renganayar, 2010, p. 12). In
South Korea, the impacts have been hard to ignore; in 2007, the government
estimated that "around 210,000 children needed treatment for internet addiction" ("Addicted?" 2011, p. 1).

But there is still a great deal of skepticism about internet addiction. LaRose,
Kim, and Peng (2011) examined patterns of online media usage with a focus
on specific internet activities and ultimately urge an understanding of this
behavior under a habitual use model. Social networking was associated with
higher degrees of deficient self-observation, but there still appeared to be a self-
regulation process in place: "Based on the current results, social networking services appear to be no more problematic, addictive, or even habitual than
others despite their widespread popularity and popular press accounts of
'Facebook addiction'" (p. 78). For those concerned about their own patterns of

use, there is an online Facebook Compulsion Quiz that will score your behavior (see Cohen, 2009; visit http://tumanov.com/quiz/).

This still leaves a very direct question about whether voluntary private corporate action is enough to address this range of concerns, and it leads into the next sections on the social and legal framework for reducing the risks of online social interaction for teens.

## Terms of Use and Private Controls

"Facebook," Grimmelmann admits, "has the most comprehensive privacy-management interface I've ever seen. Facebook users have greater technical control over the visibility of their personal information than do users of any of its major competitors" (2009, p. 1185). Yet he concludes that "Facebook is Exhibit A for the surprising ineffectiveness of technical controls" (p. 1140). The problem is the rising norm of information-sharing and the difficulties of translating nuances into clear software rules: "As soon as the technical controls get in the way of socializing, users disable and misuse them" (p. 1140). The net effect of several common law decisions on "shrinkwrap" and "clickwrap" agreements has been to guide user agreements toward creating clear and definable relationships between the user and the site (see *Specht v. Netscape, Harris v. Blockbuster*, and *Hines v. Overstock.com*). Google, for example, has created a "Privacy Dashboard," which empowers users to set information controls for every application they use within the Google environment (Terenzi, 2010).

There are also pretty clear indications that Facebook is responsive to public outcry and to these clarity concerns. When Facebook introduced the Beacon program in 2008, which shared personal information with third parties for advertising and marketing purposes, it made it an opt-out program instead of an opt-in program. The resulting backlash caused Facebook to first shift to an opt-in and then to scrap the program altogether (Hashemi, 2009; Scoble, 2010).

The most recent iteration of the privacy section of the Facebook Terms of Use was released in March of 2011 and featured a much clearer set of explanations and an even more extensive set of controls. Not too coincidentally, Facebook's announcement came as the White House was holding a conference on cyberbullying (Tapellini, 2011) to publicize the launch of a multimedia campaign for awareness about the effects of this type of bullying. Facebook provided users with greater resources to combat hostile online activity, dubbed "social reporting": "By encouraging people to seek help from friends, we hope that many of these situations can be resolved face to face…" (Pleshaw, 2011, p. 1).

Because Facebook is an "interactive computer service," with users as "content providers," it is subject to the requirements and immunities of the Communications Decency Act (Hashemi, 2009, p. 159). Functionally, this means Facebook is protected from prosecution for actions taken by third parties using the service (a finding broadly applied by the Fourth Circuit court in *Zeran v. America Online, Inc.*). More recent cases in the Seventh Circuit (*Chicago Lawyers' Committee v. Craigslist*) and the Ninth Circuit (*Fair Housing Council v. Roommates.com*) have redefined and limited the application of such immunity, leaving open the possibility that privacy violations by Facebook might be granted relief (Millier, 2009, pp. 555–556).

## "Discovering" Information on Facebook

Facebook receives 10 to 20 information requests from law enforcement every day. Some of these are court-ordered subpoenas, while others are brought under the Electronic Communications Privacy Act (ECPA) of 1986 (Chahal, 2011). A portion of the ECPA, the Stored Communications Act, allows government authorities to access time and date logs if there is "reasonable grounds to believe" that the logs are "relevant to an ongoing criminal investigation" (Chahal, 2011, p. 295 The issue is complicated since the ECPA is plagued with terminology from the state of technology in 1986, but it still manages to be more favorable to the prosecution than to defense attorneys. More recently, defense attorneys have been asking judges to order plaintiffs to sign consent forms, which get attached to subpoenas for accessing private information. In at least two cases, state courts have granted defendants "broad access to 'private' photos and comments" ("Facebook Used," 2011, p. 1), both involving personal injury claims refuted in part by photos of the parties on a fishing trip and on vacation in Florida.

For parents of teenagers who might post information that would threaten future (or current) employment opportunities, there are mixed signals. In 2010, a survey by Microsoft revealed that 75% of U.S. recruiters and human resource professionals are required by their companies to conduct online research about candidates. Almost 70% of U.S. recruiters have rejected candidates because of online information (Rosen, 2010). In the case of Stacy Snyder, who posted a picture of herself wearing a pirate hat and drinking from a plastic cup with the caption, "Drunken Pirate," she was denied her teaching degree for condoning drinking and conduct unbecoming a teacher. Because she was a public employee and the photo didn't relate to matters of public concern, the judge denied that it was protected speech (Rosen, 2010). But if employers misrepresent themselves to gain information in Facebook, that may be action-able as a violation of Facebook's Terms of Use (Brandenburg, 2008). And more recently, a company that dismissed an employee for making critical remarks about her supervisor on her Facebook site settled with the National Labor Relations Board and agreed to revise its policies about employees dis-cussing working conditions while not at work ("Company Settles," 2011).

## Social Marketing and Information Ownership

With both the Beacon and the more recent Social Ads functions, Facebook has tried to make community ties function for commercial purposes. Dubbed *social marketing*, this allows marketers to send personalized promotional messages fea-turing a teenage customer's purchases and endorsements to that customer's friends. McGeveran (2009) argues that the practice falls between the cracks of privacy protection, unfair competition, and consumer rights, and has several substantial implications for individual rights:

> First and most clearly, the individual disclosures made in social marketing messages could impinge on personal privacy. Second, some designs for social marketing could cause information quality problems that degrade not only the effectiveness or accuracy of a particular endorsement, but also the entire ecology of online peer recommenda-tion. Finally, social marketing threatens to rob the individual of control over the commercial exploitation of identity and reputation. (p. 1108)

Grimmelmann's (2009) conclusion is that the rights of publicity should apply when Facebook "turns its users into shills" (p. 1197), and that there should be a stronger consent process before user behavior or user content may be used for advertising or marketing purposes.

In general, the regulatory framework for issues confronting teenagers and their parents regarding Facebook use is patchwork at best. Unlike heightened European concern about privacy issues, and a concomitant stronger and more centralized regulating authority, the United States has not responded particularly cohesively to the challenges. In the case of social network sites, Grimmelmann (2009) notes that, "there's a deep, probably irreconcilable tension between the desire for reliable control over one's information and the desire for unplanned social interaction" (p. 1185). There may be some comfort to parents that author-ities can access records to prosecute teen endangerment, and that employers may have to tone down policies on private speech, but the array of federal, state, and local statutes, along with contradictory case law, and vested corporate interests, make outcomes highly unpredictable and therefore even a bit more fearful.

# III. Facebook and Teen Empowerment

What, then, of parental hopes for their children? Is our legacy to our youth a technology that underscores their narcissism, feeds their stimulus habit, and pro-vides a community based on a stream of social banality? Or are there truly deeper currents here that reflect a growing power and agency for rising gener-ations? This section begins with the more traditional roots of political power, explores some of the counterstatements issued by critics of social network activism, and concludes with a broader sense of empowerment.

## The New Gold Rush in Electoral Politics

When Howard Dean's campaign raised $8 million in the second quarter of 2003, politicos started paying serious attention to the internet as a political force (Haynes & Pitts, 2009). In 2008, every campaign had a web presence, and yet it was the Obama campaign that truly began to tap the power of social media. In discussing the "simulacrobama," Friedman (2009) notes that the Obama campaign employed over 90 internet staffers and raised more than half a bil-lion dollars, of which more than 92% was comprised of donations less than $100. In modern terms, Friedman explains, Obama was the sleek and sexy Macintosh, Clinton was the powerful yet more traditional PC, and McCain was the IBM Selectric.

The inspired approach of the Obama campaign was to decentralize power and enable action steps, rallies, and events to take place at the grassroots level fueled by a network of engaged young adults. The strength of social media, Metzgar and Maruggi (2009) argue, is that it is communal in nature and lacks a strict hierar-chy. Friedman also locates the power of social networking as an organizing tool that is not affiliated with a campaign and is "undisciplined" by managers (p. 347). Metzgar and Maruggi conclude that social media offer not just dissemination of information but active involvement in being part of the network.

## Clicking on the Bandwagon

The "rush to Facebook" was particularly notable in the 2010 elections. Consultants have been hired to help make Republican Congressional leaders more Facebook savvy (Shapira, 2009), and John Boehner has proclaimed that the House Republicans have "dominated" the social media contest (Farley, 2010), leading Democrats in the amount of tweeting, in Twitter followers, in YouTube channel viewers and subscribers, and in Facebook friends. Shogan, of the Congressional Research Service, tracked 2 weeks of tweeting and found that 69.8% originated from the House Republicans (2010, p. 231). The Special Counsel Office has even fashioned "rules of the road" for federal employees using social media (Davidson, 2010).

Yet young voters have not remained consistently engaged in politics. Heather Smith, of Rock the Vote, notes that the midterm elections have failed to mobilize students on campus and elsewhere (2010, p. B2), since they are disaffected by the emphasis on the Tea Party and are looking for evidence of the change they advocated for in 2008. "House, Senate and gubernatorial candidates across the country," Murray and Wagner (2010) observe, "are finding it isn't so easy to duplicate Obama's results" (p. A1).

## Social Functions of Facebook

Johnson, Zhang, Richard, and Seltzer (2011) argue that reliance on social networking sites does predict offline political participation (p. 196). They summarize research on social capital formation and explain that such sites have both a *bonding capacity* (for developing strong ties) and a *bridging capacity* (for developing weak ties with more people). Young voters, they note, use social networks for political purposes almost twice as much as older voters and see themselves as "more than consumers of news but conduits, emailing friends links and videos, and receiving them in turn" (p. 190). However, in a comparison of uses and gratifications between social networking sites and blogs, Kaye (2011) found significant differences between heavy social network users and heavy bloggers: Social networkers are significantly less interested in and perceive themselves as less knowledgeable about politics, and are more trusting of the government, but they are lower in self-efficacy than blogophiles (p. 220). This leads us again to the larger immigrant question about the real efficacy of this kind of engagement.

## Is This Really Social Activism?

There remain persistent doubts about the efficacy of "click and join" politics. Malcolm Gladwell, of *The New Yorker*, largely dismisses Facebook users' capacity to effect social change. Adams (2010) summarizes his trenchant observation:

> [W]hile social networks may be useful for some communication—to alert like-minded acquaintances to social events, or to solve a specific "weak tie" problem, such as the location of a bone marrow donor—they do not promote the passionate collective engagement that causes individuals to make commitments that result in social change. (p. 22)

In his discussion of emerging media and the "revenge of publicity," Barney (2008) makes the more darkly critical observation that while 40% of Americans went online in 2008 to get news or campaign information, 35% watched campaign-related videos online, and 29% used the internet to access primary campaign materials, only 10% used social networking sites to access political information, 5% have posted their own political views, and only 2% have used the internet to sign up for campaign-related volunteer activities. Barney (2008) argued as follows:

> Participation in these exercises serves primarily to take the edge off political appetites for justice, appetites whose satisfaction might otherwise demand more robust and meaningful forms of judgment and action, and to apply a veneer of legitimacy to governing practices that systematically oppose the more radically democratic ends of politicization and material equality. (p. 94)

Morozov (2011) offers a similar indictment of the Western media theme that Twitter and Facebook are the new Voice of America and saviors of democracy. In a social medium based primarily on narcissism, he terms many efforts on Facebook as *slacktivism* and identifies a Facebook cause of "Saving the Children of Africa," which features an impressive 1.7 million members, who have raised a total of $12,000 (less than one-hundredth of a penny per person; p. 190). In the international fight for democracy, the titular "Net Delusion" he speaks of combines cyberutopianism (set of beliefs about the inherently emancipatory nature of online communication) and net-centrism (a philosophy of action that frames every question about democratic change in terms of the internet). Authoritarian governments, he observes, have had plenty of experience using the internet in service of a tightly controlled regime. To fight authoritarianism, Morozov concludes, we need to ditch our flawed assumptions (cyberutopianism) and our flawed and crippled methodology (internet-centrism). Our best hope, he concludes, is a theory of action sensitive to the local context that "originates not in what technology allows, but in what a certain geopolitical environment requires" (p. xvii).

## Social Empowerment Writ Large and Small

While there are grains of truth in the observations about banality in the daily sharing of life stories, it is also somewhat unfair to put the failure of international action to foster democracy on the shoulders of teens and young adults. What scales in networked publics, boyd (2011) writes, "is often the funny, the crude, the embarrassing, the mean, and the bizarre" (p. 48). Yet even the politics of the personal, the banal exchange of fragmented stories, can transcend the turgid individuality and tedium of daily life. Three illustrations ground this hope in the viability of community.

First, examples of Facebook campaigns that affected some level of social change are numerous:

- A student at William & Mary organized a movement and protest against the Virginia Attorney General's attempt to get colleges and universities to remove prohibitions on discrimination on the basis of sexual orientation. Within 48 hours, more than 700 students had signed up to protest. (Unze, 2010, p. 3A).
- On Long Island, a huge housing development project was halted by a Facebook page called, "Say NO to Avalon Bay Huntington Station," sparking the real estate industry to offer seminars on social media and development. (Yan & Morris, 2010, p. A16)

On hearing of a planned protest at a high school in Woodbridge, Virginia, by the Westboro Baptist Church, parents created a counterevent and asked people to pledge money and demonstrate with signs countering the Westboro message. This frequently happens at county high schools. (Feather, 2010, n.p.). Is this world peace? Is it a victory for international democratization? Maybe not. Is it transformational dialog? Is it joining together for real action in the face of real need? Quite likely.

Second, identity formation is empowered by technology affordances, but it is also empowering within a sociopolitical context. Brabazon puts the issue starkly: "Like-minded people who share a class, literacy and technological competence can converse with people like themselves. This is not democracy. It is country club on a computer" (2007, p. 1). This is precisely why ethnic identity formation and interaction can be so powerful in Facebook. Grasmuck et al. (2009) investigated Facebook interaction among different ethnic groups and found that visual identity (photos), cultural identity (tastes and preferences), and "About Me" claims were highly elaborated and salient for several ethnoracial identity groups. Empowerment is not something personal but something that links individual strengths to a wider environment; by providing virtual "identity workshops" (Amichai-Hamburger, McKenna and Tal, 2008), Facebook allows users to reframe themselves and "reshape the terrain of human interaction," enabling the marginalized to find a voice (Grasmuck et al., 2009,

p. 180). This becomes especially potent in the sharing of images and stereotypes among diverse and oppositional cultures. Analyzing perceived stereotypes and images among Americans and Palestinians, Alhabash (2009) notes there was a positive change for both groups as a result of interacting together for a month through a closed Facebook group (p. 21). Alhabash expresses hope for a new kind of public diplomacy, where the process is not top down or bottom up but horizontal among people themselves (p. 25).

Third, there is great power in shared experience that transcends life on Earth. Wandel (2009) notes that as many as 22% to 30% of college students are in the first year of grief following the death of a friend or a family member. She explains the power of shared grief at Virginia Tech that made Facebook change its policy of deleting profiles of the deceased. Online protests and a letter campaign preserved these "virtual memory walls" of the victims of the on-campus massacre. A student compared a 2003 death in the family, where he or she received a few scattered condolence cards, with the outpouring of support that was received on Facebook after the death of the student's grandfather in 2008. Although one student expressed reservations at the impersonal nature of the interaction, another commented on the diversity of supporters, from a range of different social circles and stages of life, and concluded that the "pros of social media and bereavement outweigh any cons" (p. 48). Helping others work through powerful emotions in life does not always involve a human touch, but it is always a human gesture.

Following boyd's (2011) discussion of networked publics as simultaneously both a collective of people and a place, we may ultimately find that rather than creating homophilous social networks, as she notes, we are radically reconstructing the physicality of our culture. Richard Coyne concludes as follows:

> Place is both a source and a medium of agency, a phenomenon brought to light most potently through the incursion of ubiquitous digital devices and media. The physicality of a place is an aspect of the cultural scaffolding within which thought is constructed. Mobile phones and social media also provoke and amplify difference, and reveal further the nature of interpretative communities as decisive agents. Participative design, aided and abetted by communications technologies, is politically attuned and liberalizing. It is also more cognitively accurate as a description of the way things get done. (2009, p. 130)

## Note

1. The number of Facebook subscribers reached 835 million worldwide by March 2012, according to http://www.internetworldstats.com.

# References

Acohido, B. (2011, March 1). Sex predators stalk social media; laptops, mobile apps can make it hard for parents to monitor kids. *USA Today*, p. 1A.

Adams, T. (2010, October 3). Social media: Can Twitter really change the world? *Observer*, p. 22.

Addicted? Really? (2011, March 12). *Economist*, p. 1.

Alhabash, S. (2009, May 24). *Youth 2 youth: Changing Palestinian-American images and stereotypes through Facebook*. Paper presented at the meeting of the International Communication Association, Chicago, Illinois.

Amichai-Hamburger, Y., McKenna, K., & Tal, S. (2008). E-empowerment: Empowerment by the internet. *Computers in Human Behavior, 24*, 1776–1789.

Bahadur, R. (2009). Electronic discovery, informational privacy, Facebook and utopian civil justice. *Mississippi Law Journal, 79*, 317–369.

Baig, E., & Swartz, J. (2010, November 16). Facebook expands its messages. *USA Today*, p. B1.

Ball, A. (2010, June 20). Reflections in the Facebook mirror. *New York Times*, p. ST2.

Barney, D. (2008). Politics and emerging media: The revenge of publicity. *Global Media Journal—Canadian Edition, 1*, 89–106.

Bauerlein, M. (2008). *The dumbest generation: How the digital age stupefies young Americans and jeopardizes our future*. New York: Penguin.

Bonds-Raacke, J., & Raacke, J. (2010). Myspace and Facebook: Identifying dimensions of uses and gratifications for friend networking sites. *Individual Differences Research, 8*, 27–33.

boyd, d. (2011). Social network sites as networked publics: Affordances, dynamics, and implications. In Z. Papacharissi (Ed.), *A networked self: Identity, community, culture on social network sites* (pp. 39–58). New York: Routledge.

Brabazon, T. (2007, August 29). Enough already with the Facebook! Artshub.com. Retrieved from http://www.artshub.com.au/au/news-article/opinions/museums-and-libraries/comeback-karl-all-is-forgiven-165380

Brandenburg, C. (2008). The newest way to screen job applicants: A social networker's nightmare. *Federal Communications Law Journal, 60*, 597–615.

Brown, C., & Czerniewicz, L. (2010a). Debunking the "digital native": Beyond digital apartheid, towards digital democracy. *Journal of Computer Assisted Learning, 26*, 357–369.

Brown, C., & Czerniewicz, L. (2010b). Describing or debunking? The net generation and digital natives. *Journal of Computer Assisted Learning, 26*, 317–320.

Burhanna, K., Seeholzer, J., & Salem, Jr., J. (2009). No natives here: A focus group study of student perceptions of Web 2.0 and the academic library. *Journal of Academic Librarianship, 35*, 523–532.

Carr, N. (2010). *The shallows: What the internet is doing to our brains*. New York: Norton.

Chahal, S. (2011). Balancing the scales of justice: Undercover investigations on social networking sites. *Journal on Telecommunications & High Technology Law, 9*, 285–311.

Cohen, E. (2009). Five clues that you are addicted to Facebook. CNNHEALTH.com. Retrieved from http://www.cnn.com/2009/HEALTH/04/23/ep.facebook.addict/

Company settles case in firing tied to Facebook. (2011, February 8). *New York Times*, p. B7.

Coyne, R. (2009). Interpretative communities as decisive agents: On pervasive digital technologies. *Theory, 13*, 127–131.

Davidson, J. (2010, August 24). Guidelines on Hatch Act and social media are "rules of the road" for upcoming elections. *Washington Post*, p. B3.

Debatin, B., Lovejoy, J., Horn, A., & Hughes, B. (2009). Facebook and online privacy: Attitudes, behaviors, and unintended consequences. *Journal of Computer-Mediated Communication, 15,* 83–108.

Facebook used as legal tool. (2011, January 28). ITWeb Online. Retrieved from http://www. itweb.co.za/index.php?option=com_content&view=article&id=40557:facebook-used-as-legal-tool&catid=147

Farley, R. (2010, January 25). Boehner claims House Republicans dominate Twitter, YouTube and other social media in Congress. *St. Petersburg Times*, p. 1.

Feather, M. (2010, November 16). Virginia UUs drown out Westboro Baptist Church's hate message with love. [Blog Post]. Retrieved from http://www.standingonthesideoflove.org/blog/virginia-uus-drown-out-westboro-baptist-churchs-hate-message-with-love/

Friedman, M. (2009). Simulacrobama: The mediated election of 2008. *Journal of American Studies, 43,* 341–356.

Gelman, L. (2009). Privacy, free speech, and "blurry-edged" social networks. *Boston College Law Review, 50,* 1315–1344.

Gozzi, R. (2010, April). My life is a drama on Facebook. *ETC.: A Journal of General Semantics, 67,* 234–235.

Grasmuck, S., Martin, J., & Zhao, S. (2009). Ethno-racial identity displays on Facebook. *Journal of Computer-Mediated Communication, 15,* 158–188.

Grimmelmann, J. (2009). Saving Facebook. *Iowa Law Review, 94,* 1137–1206.

Gross, R., & Acquisti, A. (2005). *Information revelation and privacy in online social networks.* Workshop on Privacy in the Electronic Society, Carnegie Mellon University, Pittsburgh, PA.

Hargittai, E. (2010, February). Digital na(t)ives? Variation in internet skills and uses among members of the "net generation." *Sociological Inquiry, 80,* 92–113.

Hashemi, Y. (2009). Facebook's privacy policy and its third-party partnerships: Lucrativity and liability. *Boston University Journal of Science and Technology Law, 15,* 140–161.

Haynes, A., & Pitts, B. (2009, January). Making an impression: New media in the 2008 presidential nomination campaigns. *Political Science and Politics, 42,* 53–58.

Hetcher, S. (2008). User-generated content and the future of copyright: Part two—agreements between users and mega-sites. *Santa Clara Computer & High Technology Law Journal, 24,* 829–867.

Hoffman, J. (2010, December 5). Poisoned web: As bullies go digital, parents play catch-up. *New York Times*, p. A1.

Hoover, J. (2006, May 1). Beware child predators; teens hang out on social web sites—but so do sexual predators. *Information Week*, p. 34.

Jackson, M. (2009). *Distracted: The erosion of attention and the coming dark age.* Amherst, NY: Prometheus.

Johnson, T., Zhang, W., Richard, S., & Seltzer, T. (2011). United we stand? Online social network sites and civic engagement. In Z. Papacharissi (Ed.), *A networked self: Identity, community, culture on social network sites* (pp. 185–207). New York: Routledge.

Jones, T. (2008, January 10). A deadly web of deceit. *Washington Post*, p. C01.

Jukes, I., McCain, T., & Crockett, L. (2010). *Understanding the digital generation: Teaching and learning in the new digital landscape*. Thousand Oaks, CA: Corwin.

Kaye, B. (2011). Between Barack and a net place: Motivations for using social network sites and blogs for political information. In Z. Papacharissi (Ed.), *A networked self: Identity, community, culture on social network sites* (pp. 208–231). New York: Routledge.

Kennedy, G., Judd, T., Dalgarnot, B., & Waycott, J. (2010). Beyond natives and immigrants: Exploring types of net generation students. *Journal of Computer Assisted Learning, 26,* 332–343.

Khadaroo, S. (2010, October 8). Report: One-third of US teens are victims of cyberbullying. *Christian Science Monitor*, p. 1.

Lappin, Y., & Eglash, R. (2011, January 6). Facebook allegedly at center of teen suicide and rape. *Jerusalem Post*, p. 1.

LaRose, R., Kim, J., & Peng, W. (2011). Social networking: Addictive, compulsive, problematic, or just another media habit? In Z. Papacharissi (Ed.), *A networked self: Identity, community, culture on social network sites* (pp. 59–81). New York: Routledge.

Lee, F. (2008, January 22). The rough-and-tumble online universe traversed by young cybernauts. *New York Times*, p. E1.

Lenhart, A., Purcell, K., Smith, A., & Zickuhr, K. (2010). Social media and mobile internet use among teens and young adults. Pew Internet & American Life Project. Retrieved from http://pewinternet.org/Reports/2010/social-media-and-young-adults.aspx

Livingstone, S. (2008). Taking risky opportunities in youthful content creation: Teenagers' use of social networking sites for intimacy, privacy and self-expression. *New Media & Society, 10,* 393–411.

Margaryan, A., Littlejohn, A., & Vojt, G. (2011). Are digital natives a myth or reality? University students' use of digital technologies. *Computer & Education, 56,* 429–440.

McGeveran, W. (2009). Disclosure, endorsement, and identity in social marketing. *University of Illinois Law Review, 2009,* 1105–1166.

Meredith, J. (2010). Combating cyberbullying: Emphasizing education over criminalization. *Federal Communication Law Journal, 63,* 311–340.

Metzgar, E., & Maruggi, A. (2009, Spring/Summer). Social media and the 2008 U.S. presidential election. *Journal of New Communications Research, 4,* 141–165.

Millier, S. (2009). The Facebook frontier: Responding to the changing face of privacy on the internet. *Kentucky Law Journal, 97,* 541–563.

Mitchell, K., Finkelhor, D., Jones, L., & Wolak, J. (2010). Use of social networking sites in online sex crimes against minors: An examination of national incidence and means of utilization. *Journal of Adolescent Health, 47,* 183–190.

Morozov, E. (2011). *The net delusion: The dark side of internet freedom*. New York: Public Affairs.

Mui, Y., & Whoriskey, P. (2010, December 31). Facebook cements no. 1 status. *Washington Post*, p. A12.

Mundy, L. (2010, January 21). Families and the seductive screen; for beleaguered parents, limiting children's electronics time is harder than it looks. *Washington Post*, p. C01

Murray, S., & Wagner, J. (2010, October 8). Text, tweet win? Not so fast. *Washington Post*, p. A1.

Nielsen Company. (2009, June). How teens use media. Retrieved from http://blog.nielsen.com/nielsenwire/reports/nielsen_howteensusemedia_june09.pdf

Oppenheimer, T. (2003). *The flickering mind: Saving education from the false promise of technology*. New York: Random House.

Page, R. (2010). Re-examining narrativity: Small stories in status updates. *Text & Talk, 30*, 423–444.

Palfrey, J. (2010). The challenge of developing effective public policy on the use of social media by youth. *Federal Communications Law Journal, 63*, 5–17.

Palfrey, J., & Gasser, U. (2008). *Born digital: Understanding the first generation of digital natives*. New York: Basic Books.

Park, N., Kee, K., & Valenzuela, S. (2009). Being immersed in social networking environment: Facebook groups, uses and gratifications, and social outcomes. *Cyberpsychological Behavior, 12*, 729–733.

Pleshaw, G. (2011, March 10). Facebook adds more ways to report bullies. Retrieved from http://www.allfacebook.com/facebook-adds-more-ways-to-report-bullies-2011–03

Prensky, M. (2001). Digital natives, digital immigrants. *On the Horizon, 9*, 1–6.

Prensky, M. (2010). *Teaching digital natives: Partnering for real learning*. Thousand Oaks, CA: Corwin.

Press room statistics. (2011). Facebook. Retrieved from http://www.facebook.com/press/info.php?statistics

Quan-Haase, A., & Young, A. (2010, October 1). Uses and gratifications of social media: A comparison of Facebook and instant messaging. *Bulletin of Science, Technology & Society, 30*, 350–361.

Renganayar, C. (2010, August 1). Do you have Facebook addiction disorder? *New Straits Times*, p. 12.

Rosen, J. (2010, July 31). The persistent memory: Internet. *The Age*, p. 5.

Rosen, L., Carrier, M., & Cheever, N. (2010). *Rewired: Understanding the iGeneration and the way they learn*. New York: Macmillan.

Scoble, R. (2010, April 30). Why it is too late to regulate Facebook. *The Business Insider*, p. 1.

Shapira, I. (2009, December 30). Lawmakers find a friend in the power of Facebook; site lobbyist, 25, updates clients in Congress on how to leverage their accounts. *Washington Post*, p. C1.

Shogan, C. (2010, April). Blackberries, tweets, and YouTube: Technology and the future of communicating with Congress. *Political Science & Politics, 43*, 231–233.

Small, G., & Vorgan, G. (2008). *iBrain: Surviving the technological alteration of the modern mind*. New York: Collins Living.

Smith, D. (2008, March 23). Addiction to internet "is an illness": Heavy users suffer isolation, fatigue and withdrawal symptoms, says a psychiatrist. *Observer*, p. 11.

Smith, H. (2010, October 24). 5 myths about young voters. *Washington Post*, p. B2.

Smith, S., & Caruso, J. (2010). *The ECAR Study of Undergraduate Students and Information Technology, 2010*. Boulder, CO: Educause.

Tapellini, D. (2011, March 11). White House conference aims to combat cyberbullying. ConsumerReports.org. Retrieved from http://blogs.consumerreports.org/electronics/2011/03/white-house-conference-aims-to-combat-cyberbullying.html

Tapscott, D. (2009). *Grown up digital: How the net generation is changing your world*. New York: McGraw-Hill.

Taraszow, T., Aristodemou, E., Shitta, G., Laouris, Y., & Arsoy, A. (2010). Disclosure of personal and contact information by young people in social networking sites: An analysis using Facebook profiles as an example. *International Journal of Media and Cultural Politics, 6,* 81–101.

Terenzi, R. (2010). Friending privacy: Toward self-regulation of second generation social networks. *Fordham Intellectual Property, Media & Entertainment Law Journal, 20,* 1049–1106.

Unze, D. (2010, March 26). Facebook helps movements ignite; social networking site has shown "huge impact" in grass-roots organizing. *USA Today,* p. 3A.

Valenzuela, S., Park, N., & Kee, K. (2009). Is there social capital in a social network site?: Facebook use and college students' life satisfaction, trust, and participation. *Journal of Computer-Mediated Communication, 14,* 875–901.

Wandel, T. (2009, Fall/Winter). Online empathy: Communicating via Facebook to bereaved college students. *Journal of New Communication Research, 4,* 42–52.

Wang, J., Nansel, T., & Iannotti, R. (2011). Cyber and traditional bullying: Differential association with depression. *Journal of Adolescent Health, 48,* 415–417.

Wolak, J., Finkelhor, D., Mitchell, K., & Ybarra, M. (2008, February-March). Online "predators" and their victims: Myths, realities, and implications for prevention and treatment. *American Psychologist, 63,* 111–128.

Worden, A. (2008, May 9). Facebook reaches agreement on sex predators. *Philadelphia Inquirer,* p. B01.

Yan, E., & Morris, D. (2010, September 30). Avalon defeated by social media; Facebook pages had big role in fate of controversial project; influence called eye-opener for LI's real estate industry. *Newsday,* p. A16.

# · 3 ·

# Swimming in Cyber-Cesspools

## Defamation Law in the Age of Social Media

DALE A. HERBECK

JuicyCampus launched on seven college campuses—Duke University, Loyola Marymount University, UNC, USC, Pepperdine, UCLA, and the College of Charleston—in August 2007 and quickly spread to other institutions. By March 2008, the increasingly controversial site had expanded to 59 institutions. By October 2008, JuicyCampus had reached more than 500 campuses and the site boasted more than one million unique visitors every month. Operating under the slogan, "C'mon. Give us the juice," the web site held itself out as the "place to spill the juice about all the crazy stuff going on at your campus." Since JuicyCampus guaranteed anonymity to posters, the discussion quickly degenerated into outrageous threads about the "most overrated person," the "hottest freshman girl," "guys with STDs," and the "cutest non slutty sorority girl."

To the delight of its critics, JuicyCampus abruptly announced it was suspending operations on February 4, 2009. By all accounts, the controversial web site was done in by the failure of its business model. Like more traditional web sites, JuicyCampus had hoped to sustain itself with a revenue stream generated from advertising. "Even with great traffic and strong user loyalty," the web site lamented, "a business can't survive and grow without a steady stream of revenue to support it" (Official JuicyCampus Blog, 2009). Although JuicyCampus attributed its demise to "historically difficult economic times," the site failed because advertisers were unwilling to associate either their corporate name or their products with a web site that had become notorious for its offensive content.

The death of JuicyCampus does not mean, however, that there is no place for gossip or rumors in cyberspace. To the contrary, the explosive growth of the social media has created a dizzying array of *cyber-cesspools*, a term coined by Brian Leiter (2010) to refer to "places in cyberspace—chat rooms, websites, blogs, and other comments sections of blogs—which are devoted in whole or in part to demeaning, harassing, and humiliating individuals; in short, to violating their 'dignity'" (p. 155). "JuicyCampus," Daniel Solove observed, "is really just an exclamation point following everything else that's already going on" (as cited in Morgan, 2008). This infamous site is noteworthy only because of the aggressive and unabashed way that it solicited salacious content. Even without JuicyCampus, there are a myriad of ways in which social media have been used to damage someone's reputation. Consider three such possibilities, each illustrated with an actual example that had real consequences for the unfortunate target:

1. Wikis. It is possible to post false or outrageous statements on wikis, collaborative web sites that allow users to add, remove, and change content. This is exactly what happened to John Seigenthaler, Sr., when someone altered his biography on Wikipedia in 2005 to include the following claim: "John Seigenthaler Sr. was the assistant to Attorney General Robert Kennedy in the early 1960's. For a brief time, he was thought to have been directly involved in the Kennedy assassinations of both John, and his brother, Bobby. Nothing was ever proven." Seigenthaler, who had served as Robert Kennedy's administrative assistant and was a pallbearer at his funeral, was understandably outraged when he learned (132 days after the content was posted) of the accusation that he had played a role in either of the assassinations (Seigenthaler, 2005, p. 11A).

2. Bulletin boards. It is possible to anonymously make disparaging comments on computer bulletin boards, forums devoted to discussing different topics. One such forum is AutoAdmit, an unmoderated board for prospective and current undergraduate and graduate students that draws between 800,000 and one million viewers per month. AutoAdmit drew national attention when anonymous posters—using names like AK47, stanfordtroll, and Dirty Nigger—targeted two female students attending Yale Law School in 2007. Postings claimed that one of the women had bribed her way into Yale and had a sexual affair with an administrator. It

was alleged the other woman had gonorrhea and was addicted to heroin. If this were not disturbing enough, the posts also featured pictures of both students and threatened criminal and sexual violence against them (*Doe I and Doe II*, 2008, p. 251).

3.  Myspace or Facebook. It is possible to create fake profiles using a social networking site such as Myspace. Angry at being disciplined for dress code violations, two Pennsylvania middle school students created a fictitious Myspace profile for James McGonigle, the principal of Blue Mountain Middle School in Orwigsburg, Pennsylvania, on March 18, 2007. Although the students did not refer to their principal by name, they used McGonigle's photograph, which they obtained from the school district's web site. To complete the profile, they answered a series of questions provided by Myspace. The page that resulted contained vulgar and crude language and purported to be about "M-Hoe," a 40-year-old, married, bisexual man living in Alabama. The profile listed the following interests: "detention, being a tight ass, riding the fraintrain, spending time with my child (who looks like a gorilla), baseball, my golden pen, fucking in my office, hitting on students and their parents" (*J.S. v. Blue Mountain School District*, 2011, p. 4).

These are, unfortunately, not isolated episodes as there are a growing number of instances where the social media were wielded as a potent weapon to damage someone's reputation. This is an old problem, but it is magnified by new technology because of the vast number of users, the power of search engines like Google, and the permanence of digital information.

"Rumors are nearly as old as human history," Cass Sunstein (2009) writes, "but with the rise of the Internet, they have become ubiquitous. In fact we are now awash in them" (p. 3). The new social media have amplified an old problem by expanding the number of people with ready access to the means of mass communication. Further compounding matters, this sort of user-generated content is instantaneous, meaning speakers may exercise less forethought before publishing material. The potential audience is vast, as is demonstrated by the fact that there were more than 800 million active Facebook users in July 2011. While the total number of users is impressive, a staggering 500 million users log onto Facebook every 24 hours.

Search engines like Google, which allow users to efficiently explore the vastness of cyberspace for content, also contribute to the problem (Pasquale, 2010). Because these powerful search engines have the capacity to search the complete text of millions of documents, a disparaging comment buried in an obscure blog can appear at the top of a search results list. "A snarky insult, embedded in a story or a post, quickly gets traffic," David Denby (2009) observes. "It gets linked to other blogs and soon it has spread like a sneezy cold through the vast kindergarten of the Web" (Denby, 2009, p. 10). While Google's search algorithm is a closely guarded secret, the number of links associated with a particular page influences search results. Recognizing that fact, a variety of attempts have been made to "shape" Google results to serve a particular agenda (Grimmelmann, 2008/2009, p. 939). So, for example, a group of Democratic activists launched a campaign to link George W. Bush's official biography page to the phrase "miserable failure" in 2003. By adding this phrase to pages that link to the official Bush page, the "Googlebombers" were able to trick the search engine into making a connection. While this is an extreme example, it neatly illustrates why "Google is part of the problem" (Leiter, 2010, p. 161).

Looking beyond the way in which Internet scales and the search engines compile results, there is also a growing awareness that communication using social media is surprisingly durable. Unlike paper, which can be destroyed, or speeches, which can be forgotten, content on the social media is collected and often available in public archives. Digitization, combined with cheap storage and easy retrieval, has created the infrastructure necessary to preserve staggering amounts of information. Youthful indiscretions, intemperate diatribes, and scurrilous accusations can live forever in cyberspace. For the first time in human history, Viktor Mayer-Schönberger (2009) writes, "remembering [is] cheaper than forgetting" (p. 197). In the new era of "persistent memory," it is unlikely that the passage of time alone will heal emotional wounds or restore a damaged reputation" (Citron, 2010, p. 1813).

According to Terence Lau (2011), the "three characteristics that make the Internet so appealing—its speed, reach and permanence—also make it a devastatingly effective tool to ruin the lives of innocent citizens" (p. 253). Traditional legal remedies, such as filing a defamation lawsuit, may offer some hope of obtaining redress in extreme cases. When characterizing the speech that permeates some of the more notorious cyber-cesspools, Leiter (2010) has drawn a distinction between *tortious harms* and *dignitary harms*. While dignitary harms are quite real to the aggrieved parties, they are not actionable because the speech is deemed worthy of constitutional protection. At some point, however, dignitary harms become substantive enough to transform into tortious harms and give rise to causes of action for "defamation and infliction of emotional distress" (p. 155). When this happens, the aggrieved party might obtain meaningful legal redress.

Despite claims that cyberspace transcends traditional legal constructs, the courts have not crafted different rules for assessing defamatory speech spread using the new social media. Without exception, the basic elements of defamation apply whether the statement was made in a public speech broadcast over the public airwaves, printed in a newspaper, or posted to JuicyCampus. There are, however, three questions worthy of consideration: (a) Is speech on the new social media entitled to the same First Amendment protection as speech on the traditional media? (b) Can an Internet service provider or an entity that created a forum (such as JuicyCampus or Twitter) be held legally responsible for defamatory content created by a third party? (c) Under what circumstances can an aggrieved party learn the identity of an anonymous speaker? By attempting to answer these questions, this analysis explores some of the new issues raised by defamatory speech in the social media.

## Traditional Media Versus Social Media

When he launched Whole Earth 'Lectric Link (The WELL), Stewart Brand coined the aphorism YOYOW, or "You Own Your Own Words" (Hafner, 2001, p. 11), that greeted members whenever they logged on to the WELL. This aphorism plays on two different meanings of the word "own." One sense suggests that speakers "own" the rights to their own speech. So, for example, copyright law allows individuals to control original intellectual property that they have created. Thus, users "own" the rights to their expression. In another sense, speakers also "own" the legal responsibility for their speech. Besides owning the rights to profit from their intellectual property, they are also culpable for any injury caused by their expression. Applied to defamation, the YOYOW assigns responsibility to the speaker. If you make a defamatory posting to a computer bulletin board or put a libelous statement on a web page, the person you defame might sue you for damages. Depending on whether your target was a public figure or a private person, the appropriate standard for proving fault could be deduced and the case would proceed like a traditional defamation case.

The potential for extending defamation law to the social media can be illustrated using a contemporary example. It is widely believed that the first Twitter lawsuit involved intemperate tweets by Courtney Love, a notorious rock star and former lead singer of the alternative rock band Hole. Like many cases, the controversy started as a trivial matter: a dispute over a bill for services. Love had commissioned fashion designer Dawn Simorangkir to transform some of her old clothing into stylish new outfits. Simorangkir, who designs the Boudoir

Queen clothing line, is best known for intricate corsets and 1920s-style frocks. Since Love was a celebrity, she apparently thought Simorangkir would provide her design services for free. Love was, therefore, offended when she received a bill for $4,000. When Love ignored a second bill, Simorangkir suspended work on Love's clothes. This sent Love over the edge and, unable to contain herself, she issued no fewer than 10 allegedly defamatory tweets in a 21-minute rant on March 17, 2009. Among other things, Love tweeted that Simorangkir "has a history of dealing cocaine, lost all custody of her child, [and was charged with] assault and burglary"(Complaint, *Simorangkir*, 2009, p. 9). Just for good measure, Love also referred to Simorangkir as a "nasty lying hose bag thief" who "gets to haul her 52 year old desperate cokes out ass to jail where they don't have three bottles of vodka a night" (Complaint, *Simorangkir*, 2009, p. 9). At the time she made the tweets, Love had more than 40,000 Twitter followers. To make sure everyone knew about the Boudoir Queen, Love also made disparaging comments about Simorangkir on Myspace and on Etsy.com, an online marketplace used by many independent designers.

Simorangkir responded to the verbal assault by suing Love for defamation, false light invasion of privacy, intentional infliction of emotional distress, and breach of contract. "Whether caused by a drug induced psychosis, a warped understanding of reality, or the belief that her money and fame allow her to disregard the law," the complaint charged, "Love has embarked in what is nothing short of an obsessive and delusional crusade to terrorize and destroy Simorangkir" (Complaint, *Simorangkir*, 2009, p. 2). Love moved for dismissal under California's anti-SLAPP law, a measure designed to limit strategic lawsuits against public participation. In October 2009, a state judge denied Love's motion on the grounds that the tweets did not deal with a matter of public concern. Further, the court held Simorangkir had demonstrated a high probability of prevailing in her defamation action. The case was scheduled for trial in January 2011 and drew considerable media attention (McCarthy, 2011). Just days before the trial commenced, a financial settlement was reached and Love reportedly paid $430,000 in damages (Belloni, 2011).

While *Simorangkir v. Love* never went before a jury, the case is notable because it suggests that the standards created for assessing speech by and in the traditional media may be extended to the social media. This might be appropriate in the case of a celebrity like Love, who had thousands of followers, and a well-known designer like Simorangkir. It is less clear, however, whether these standards should be applied to noncelebrities for statements made on social media such as Twitter or Facebook. Consider, for example, the case of

Amanda Bonnen, an angry tenant who had the audacity to tweet about conditions in her apartment building. "Who said sleeping in a moldy apartment was bad for you?" Bonnen asked."Horizon Realty thinks it's ok (Complaint, *Horizon Group*, 2009, p. 2). Although Bonnen's account was available to any Twitter user, her Google-cache revealed Bonnen had a modest 17 followers (Wright, 2009). This did not matter to Horizon Realty, which promptly filed a $50,000 lawsuit claiming the tweet was "false and defamatory" (Complaint, *Horizon Group*, 2009, p. 3). Fortunately for Bonnen, a sympathetic judge dismissed the case on the grounds that Bonnen's tweet was too vague to be defamatory.

Cases like *Horizon Realty Group v. Bonnen* illustrate the "Streisand effect." This phenomenon, which is often attributed to a 2003 incident involving Barbra Streisand, suggests efforts to avoid publicity often generate publicity. In this instance, Streisand was annoyed when Kenneth Adelman, an environmental activist, included an aerial photograph of her Malibu beach house in a collection of 12,000 coastline photos documenting coastal erosion he posted on the Internet. Streisand sued in an effort to have the photo removed, but her lawsuit was dismissed with predictable results. The Associated Press reported on her efforts and newspapers around the world reprinted the beach house photo. News of the incident translated into more than one million visitors to the web site and produced the very publicity the lawsuit intended to avoid. Mike Masnick (2005) coined the phrase the *Streisand Effect* in a commentary that ended with a simple question: "How long is it going to take before lawyers realize that the simple act of trying to repress something they don't like online is likely to make it so that something that most people would *never, ever see*…is now seen by many many more people?"

Had Horizon Realty simply ignored the tweet, only a handful of Bonnen's followers would have known about the deplorable conditions in her apartment building. By filing a defamation suit, however, Horizon called public attention to the allegation and cast itself in the role of a corporate bully. Although this case might serve as a cautionary tale and deter other aggrieved parties contemplating a defamation action, it is more likely the courts will be asked to resolve a growing number of lawsuits involving defamatory content disseminated through the social media. As the number of suits increases, courts will be forced to consider whether standards developed for the traditional media—defined here as newspapers, magazines, radio, and television—should be extended to the new social media. Since these standards are generally sympathetic to media defendants, it would be difficult for aggrieved parties to win defamation actions. As a result, the average person defamed by social media would have little chance of redress.

There have already been calls to address the problem of defamatory speech on social media. Some of these efforts—imposing intermediary liability and decreasing anonymity—are discussed in the sections that follow. It would be possible, however, to fashion different standards for private figure plaintiffs defamed by the traditional media and the social media. Along the same lines, it would also be possible to distinguish between content dealing with matters of public concern (such as Bonnen's unsafe apartment) and personal content (such as Love's unsubstantiated assertions of criminal activity). Although it is appealing, courts should resist the temptation to weaken the standards and make it easier for aggrieved parties to recover damages for defamatory content dealing with private matters disseminated using the social media. While much might be said on this point, suffice it to say that there are good reasons for tolerating some regrettable content. One of the reasons that the various social media have become so significant is that they allow different people—not just those who work for mass media organizations—to communicate. Opening the social media to large numbers of people guarantees a vibrant forum addressing a broad range of issues. The downside, of course, is that a percentage of the resulting speech will be defamatory, crude, and tasteless.

Efforts to weed out this bad expression are destined to fail. Justice John Marshall Harlan's line, "one man's vulgarity is another's lyric," sums up the impossibility of developing standards that are not hopelessly vague and subjective (*Cohen v. California*, 1971, p. 25). At the same time, making it easier for aggrieved parties to sue for defamation would necessarily chill the free speech of those who are not members of the traditional media or who seek to speak on matters of personal—but not public—interest. These are important voices and they should not be suppressed. So too, it would be a mistake to narrowly limit First Amendment protection to speech that deals solely with matters of public interest.

## Publishers Versus Distributors

Many of the early Internet defamation suits were not brought against individuals but rather Internet Service Providers (ISPs) such as America Online (AOL). ISPs are attractive targets for several reasons. First, many defamatory postings are made anonymously, so suing the ISP is often the only possible legal action. In addition, by targeting the ISP, the aggrieved party can pressure it to ascertain or even divulge the speaker's identity. Second, the ISP is often named a party to a suit, as it was the disseminator of the defamatory message. Although the ISP might not have known about the defamation, it provided Internet access and a network for distributing the message. Holding ISPs legally responsible would force them to monitor content and remove defamatory speech.

## Dueling Precedents

Two of the first cases to consider an ISP's liability for defamatory messages had very different results. In *Cubby v. CompuServe* (1991), a federal court considered CompuServe's culpability for defamatory statements posted on its "Journalism Forum" bulletin board. One part of the Journalism Forum, "Rumorville, USA," provided news and gossip about the entertainment industry. When Cubby, Inc., tried to start a rival database called "Skuttlebut," the suit alleged, "Rumorville, USA," disparaged Cubby. The allegedly defamatory remarks included the suggestion that individuals at Skuttlebut gained access to information first published by Rumorville "through some back door," a claim that Skuttlebut's founder was "bounced" by his previous employer, and a description of Skuttlebut as a "new start-up scam" (p. 138).

In issuing a summary judgment for CompuServe, the court held the following:

> First Amendment guarantees have long been recognized as protecting distributors of publications....Obviously, the national distributor of hundreds of periodicals has no duty to monitor each issue of every periodical it distributes....CompuServe has no more editorial control over...publication than does a public library, book store, or newsstand, and it would be no more feasible for CompuServe to examine every publication it carries for potentially defamatory statements than it would be for any other distributor to do so. (p. 140)

Since CompuServe was only acting as a "news distributor," circulating information created by a third party, the court granted summary judgment in favor of the ISP (*Cubby v. CompuServe*, 1991, p. 141). This did not mean there was no defamation; rather, the court held that the responsible party was the author of the content.

A New York state court came to a very different decision in *Stratton Oakmont v. Prodigy* (1995), a case involving a posting about Stratton Oakmont, a Long Island securities firm, made on the "Money Talk" bulletin board. An anonymous user alleged that Stratton Oakmont and its president, Daniel Porush, "committed criminal and fraudulent acts in connection with the initial public offering of stock of Solomon-Page Ltd" (p. 2). In response, Stratton Oakmont and Porush brought a defamation action in New York state court.

Citing the decision in *Cubby v. CompuServe* (1991) as a precedent, Prodigy filed a motion for summary judgment, arguing that it was merely distributing content created by an anonymous third party.

To the surprise of many commentators, the judge refused this request because Prodigy had exercised a measure of editorial control. In the pivotal passage, the court distinguished between the services provided by the two ISPs:

> The key distinction between CompuServe and Prodigy is twofold. First, Prodigy held itself out to the public and its members as controlling the content of its computer bulletin boards. Second, Prodigy implemented this control through its automatic software screening program, and the Guidelines which Board Leaders are required to enforce. By actively utilizing technology and manpower to delete notes from its computer bulletin boards on the basis of offensiveness and "bad taste," for example, Prodigy is clearly making decisions as to content and such decisions constitute editorial control. (p. 10)

Since Prodigy had functioned as a publisher, the court denied the motion for a summary judgment. Rather than risk a trial, Prodigy quickly negotiated an out-of-court settlement with Stratton Oakmont. An apology was issued and the lawsuit was withdrawn.

At first glance, it may seem that *Cubby v. CompuServe* (1991) and *Stratton Oakmont v. Prodigy* (1995) came to contradictory results. Under closer examination, however, the decisions highlight the important distinction between a publisher and a distributor. Under the common law, a publisher bears the same responsibility for defamatory content as the author because the publisher has the knowledge, opportunity, and ability to exercise editorial control. Because it chose to exercise this power, Prodigy functioned like the publisher of "Money Talk" and was responsible for defamatory content. In contrast, AOL exercised no editorial control and, therefore, functioned like a distributor and was not responsible for defamatory content appearing on "Rumorville, USA." Under the common law, distributors are not responsible for content they distribute because it would be impossible for a distributor such as a newsstand or a bookstore to review each and every publication on its shelves.

The message to ISPs was clear and unambiguous: If you use software to screen content or employ board leaders to police postings, you risk transforming yourself into a publisher and assuming liability for defamatory content. An ISP that made no effort to edit user-submitted content was, ironically, in a stronger legal position in a defamation lawsuit than an ISP that aggressively tried to control or edit content.

## Section 230

The *Cubby v. CompuServe* (1991) and *Stratton Oakmont v. Prodigy* (1995) decisions were a consideration when Congress debated the Communications Decency Act (CDA) of 1996. If, as some legislators wanted, an ISP actively screened content for sexually explicit content, plaintiffs in a defamation action might claim the ISP had become a "publisher" and is therefore liable for the content of defamatory postings. Fearing the *Stratton Oakmont v. Prodigy* decision would discourage ISPs from deleting sexually explicit content, Congress included Section 230 (1996), an amendment which stated, "No provider or user of an interactive computer service shall be treated as the publisher or speaker of any information provided by another information content provider." While the CDA's provisions dealing with sexually explicit speech were ruled unconstitutional in *Reno v. ACLU* (1997), that decision did not invalidate other CDA provisions or the larger bill, the Telecommunications Act of 1996, to which the CDA was attached.

The significance of Section 230 became evident in *Zeran v. America Online* (1998), a case that involved a private person residing in Seattle, Washington. The unfortunate incident began on April 25, 1995, when someone anonymously, and without the plaintiff Kenneth Zeran's knowledge or authority, affixed Zeran's first name and telephone number to postings on AOL's bulletin boards advertising T-shirts glorifying the bombing of the Alfred P. Murrah Federal Building in Oklahoma City six days earlier. After receiving threatening phone calls from outraged people and determining that the source was an AOL bulletin board, Zeran notified AOL, which removed the offending messages and closed the account from which they originated. On April 26, the unknown person posted another message announcing the T-shirts from the prior day were sold out and that new slogans were now available. This posting announced that one dollar from every sale would be donated to victims of the bombing and it also included Zeran's first name and telephone number. Over the next four days, additional messages were posted advertising additional items, including bumper stickers and key chains. By April 30, Zeran was receiving a phone call every 2 minutes, and the problem only grew worse when an announcer on Oklahoma City radio station, KRXO, learned of the postings and urged listeners to call Zeran. It was not until May 14, after an Oklahoma City newspaper exposed the hoax and KRXO made an on-air apology, that the number of phone calls finally subsided.

In January 1996, Zeran brought a defamation suit against AOL. Although he admitted that AOL did not publish the messages, Zeran sought to hold AOL responsible as a distributor for failing to promptly remove the defamatory post-ings, for not notifying users that the messages were false, and for failing to screen for future postings. AOL did not challenge Zeran's allegations of distributor lia-bility, arguing instead that Section 230 of the CDA preempted his claim. A fed-eral district court agreed. The Fourth Circuit Court of Appeals upheld the ruling for AOL, arguing, "Section 230 forbids the imposition of publisher liability on a service provider for the exercise of its editorial and self-regulatory functions" (p. 331). The Supreme Court denied certiorari.

Over the years since the *Zeran* decision, there have been dozens of cases involving Section 230. With several notable exceptions, the courts have gen-erally held that Section 230 bars suits, under federal or state law, which might impose liability on ISPs. In particular, an ISP is generally entitled to have a defamation claim dismissed if three conditions are satisfied:

1. The defendant is "a provider or user of an interactive computer ser-vice."
2. The defendant is being "treated as the publisher or speaker" of information for purposes of liability.
3. The challenged information is "information provided by another information content provider." (Ardia, 2010, p. 412).

Section 230, simply put, provides broad protection for defendants who serve as intermediaries. So long as the ISP was not the source of the defamatory con-tent, Section 230 makes it extremely difficult for an aggrieved party to success-fully sue the ISP. If the defamatory content was posted anonymously, an aggrieved party like Zeran may be left without any legal remedy.

This level of immunity has never been provided to newspapers, maga-zines, or broadcast stations. While a publisher or broadcaster is liable for defam-atory speech it publishes or broadcasts, Leiter (2010) has observed that the "owners of cyber-cesspools are held legally unaccountable for even the most noxious material on their sites, even when put on notice as to its potentially tortious nature" (p. 156). This disparity is so extreme that some critics have asked whether legislators and courts have gone too far in protecting rumors, gos-sip, and innuendo in cyberspace. Solove (2007) has complained:

> Courts are interpreting Section 230 so broadly as to provide too much immunity, elim-inating the incentive to foster a balance between speech and privacy. The way courts are using Section 230 exalts free speech to the detriment of privacy and reputation. As a result, a host of websites have arisen that encourage others to post gossip and rumors as well as to engage in online shaming. (p. 159)

Foremost among these are sites like JuicyCampus, which have the audacity to "flaunt" their protected status under Section 230 (Solove, 2010, p. 24).

Not surprisingly, there have been calls for repealing Section 230, or at least narrowing the protection to allow for notice-based liability. "Under this approach, modeled on the copyright provisions of the Digital Millennium Copyright Act," Cass Sunstein (2009) explains, "those who run websites would be obliged to take down falsehoods upon notice" (pp. 78–79). If a site like JuicyCampus or AutoAdmit failed to respond to a takedown notice submitted by an aggrieved party in a timely fashion, the operator would become liable for the defamatory content. This would create a powerful incentive for removing allegedly defamatory content and might also reduce the number of forums available for such low value expression.

At face value, there is certain intuitive appeal to such suggestions, as Section 230 often leads to tough results. At the same time, imposing a measure of liability, even notice-based liability, would inevitably spillover to reach some protected speech on the social media. Compounding matters, overly sensitive parties might use notice and takedown to target legitimate criticism. Even if the takedown notice is dubious, the operators of a web site might be tempted to remove the content to avoid the nuisance of litigation or to limit their legal liability (Seltzer, 2010, pp. 199–218). There is, after all, no downside to removing the allegedly defamatory content. Proponents of this scheme like Solove (2010) attempt to discount the risk of excessive takedowns (p. 26), but the Digital Millennium Copyright Act experience suggests there is a real potential for abuse. To punctuate this point, the Electronic Frontier Foundation maintains a "Takedown Hall of Shame," a web site that documents some of the most embarrassing content removal requests. Limiting immunity under Section 230 will, therefore, have consequences. "Whether the Internet remains open to diverse forms of speech," David Ardia (2010) concludes, "will depend to a considerable degree on whether the intermediaries that make it function can continue to be agnostic as to the content they intermediate" (p. 494).

In the final analysis, the question might hinge on the value assigned to forums like JuicyCampus and AutoAdmit. Absent some form of Section 230 protection, these controversial forums would likely disappear. Imposing notice-based liability would allow the sites to exist, but only so long as they responded to takedown notices. This is not a problem for some of the critics of Section 230. Leiter (2010) concludes that there is not enough valuable speech in and around cyber-cesspools to warrant safe harbor protection. He also believes that it would be possible to distinguish between a modern-day H. L. Mencken, who legitimately excoriates a public figure, and low-value expression, such as defamatory accusations and baseless innuendo (p. 172).

The three examples offered at the beginning of this analysis might seem to prove the point. The allegation that Seigenthaler played a role in either of the Kennedys' assassinations is false. The postings on AutoAdmit about the female law students were both offensive and threatening. And there is not much to commend the fictitious Myspace profile created by two teenage girls. These are, however, easy targets offered to illustrate how the social media can be used for mischief. It would be possible to offer an equal number of examples where the social media exposed excesses or uncovered criminal activity. Comparing examples is not, in the final analysis, the best way to make public policy. It is also unnecessary, as First Amendment jurisprudence holds that free speech produces more benefits than the alternative.

## Reputation Versus Privacy

In *McIntyre v. Ohio Elections Commission* (1995), the Supreme Court held that a state law prohibiting the distribution of anonymous campaign literature abridged the freedom of speech guaranteed by the First Amendment. The impact of the decision was muted, however, because it was difficult to reach a large audience without divulging the identity of the speaker. All of that changed with the advent of new social media that allow users to communicate through web sites, blogs, tweets, and message boards. Because of cursory registration requirements, pseudonymous remailers and other masking software, it is relatively easy for anyone to speak anonymously in cyberspace. Further compounding matters, the anonymity of the Internet may actually encourage outrageous speech. Solove (2007) observed the following:

> When people are less accountable for their conduct, they are more likely to engage in unsavory acts. When anonymous, people are often much nastier and more uncivil in their speech. It is easier to say harmful things about others when we don't have to take responsibility. (p. 140).

The current issue before the courts is not whether anonymous speech is protected, but rather, whether or not third parties who facilitate such communication should be compelled to reveal the identity of an anonymous speaker. As previously noted, it is difficult for an aggrieved party to seek redress, as Section 230 of the CDA states ISPs cannot be held responsible for third-party content. Under the YOYOW principle, the legally responsible party is the person who actually created the defamatory content. Before a lawsuit can be filed, however, it is necessary to learn the identity of the speaker. Since the content-facilitator has promised confidentiality, a growing number of lawsuits have been filed in which courts have been asked to order an ISP, a web site, or a blog to reveal the identity—or the Internet Protocol (IP) address identifying the computer where the message originated—of the anonymous speaker. Using this information, it would be possible for the aggrieved party to track down the responsible party and file a defamation claim.

State courts across the country have attempted to develop standards for assessing such claims. One influential test emerged from a 2001 New Jersey appellate court decision in *Dendrite International v. John Doe No. 3* (2001). In this case, Dendrite sought to learn the identity of four John Does who had posted statements about the company on a financial bulletin board maintained by Yahoo! In these posts, the John Does alleged Dendrite had manipulated accounting practices to artificially inflate the company's earnings. Some of the posters did not object to being identified, but John Doe No. 3 went to court to protect his anonymity. To determine whether or not to unmask the identity of an anonymous speaker, the appellate court fashioned a four-part test:

1. The plaintiff must "undertake efforts to notify the anonymous posters that they are the subject of a subpoena or application for an order of disclosure" (p. 141).
2. The plaintiff must "identify and set forth the exact statements purportedly made by each anonymous poster that plaintiff alleges constitutes actionable speech" (p. 141).
3. The plaintiff must "set forth a prima facie cause of action against the fictitiously-named anonymous defendants" (p. 141).
4. The court must "balance the defendant's First Amendment right of anonymous free speech against the strength of the prima facie case presented and the necessity for the disclosure of the anonymous defendant's identity to allow the plaintiff to properly proceed" (p. 142).

In this instance, the court refused to order Yahoo! to divulge the identity of John Doe No 3. While the company had satisfied the first two elements, the court ruled Dendrite failed the third element because the company had not made a prima facie cause of action for defamation. More specifically, Dendrite had not demonstrated an actionable injury because the anonymous postings had neither depressed the value of the company's stock nor inhibited hiring.

Subsequent decisions have adopted variations on the *Dendrite International v. John Doe No. 3* guidelines. In *Doe v. Cahill* (2005), the Supreme Court of Delaware considered a request to unmask an anonymous blogger. Writing under the alias "Proud Citizen," someone had denounced Patrick Cahill's performance as a Smyrna city councilman, charged he was "paranoid," and concluded that he was suffering from "an obvious mental deterioration" (p. 454). To facilitate his defamation suit, an outraged Cahill asked the court to order the *Delaware State News*, the operator of the blog, to disclose the poster's identity. In resolving the case, the Supreme Court of Delaware described the *Dendrite International v. John Doe No. 3* guidelines (p. 460), which it then simplified. More specifically, the court embraced the first (notification) and third (prima facie case) elements. The second element (exact statements) was deemed unnecessary and the fourth element (balancing) was subsumed in the court's decision-making process (p. 461). Since no reasonable person would interpret the statements as being factual claims about Cahill, the Supreme Court of Delaware issued summary judgment in favor of Doe.

Not all states have established standards for unmasking the identity of anonymous speakers. The state courts that have considered the question have generally embraced the *Dendrite International v. John Doe No. 3* guidelines, the *Doe v. Cahill* guidelines, or some combination of the two. At a minimum, would-be plaintiffs in a civil suit must attempt to notify the anonymous speaker whose identity is being sought, and they must demonstrate that their defamation claim has genuine legal merit before a judge will unmask an anonymous speaker. (The courts have been much more willing to order the release of an identity in a criminal case.)

This is a heavy burden, but it is not insurmountable. The court did not unmask John Doe in either *Dendrite* or *Cahill*, but not all anonymous speakers have been so fortunate. In 2009, for example, someone signed up for a free account with Google Blogger and created a blog titled, "Skanks in NYC." From this platform, the anonymous blogger made disparaging comments about Liskula Cohen, a fashion model who had previously appeared on the cover of *Vogue*. Among other things, the blog referred to Cohen as a "skank," a "ho," and "an old hag" who "may have been hot ten years ago." When Google refused to reveal who was responsible, Cohen asked a New York state judge to order Google to unmask the anonymous blogger so she could sue for defamation. Because Cohen had presented a "meritorious cause of action," the judge in *Cohen v. Google* (2009) ruled in her favor and directed Google to divulge the identity of the anonymous blogger (p. 948).Google complied and informed Cohen that

Rosemary Port, a female acquaintance who had been offended by something that Cohen had said about her to an ex-boyfriend, had registered the blog. After learning the blogger's name and confronting Port, Cohen decided not to pursue her defamation lawsuit. The "Case of the Blond Model and Malicious Blogger" serves as a cautionary tale, however, to anyone who might be inclined to use anonymity to make defamatory comments (Dowd, 2009, p. A23).

The stakes are significant. While the total number of speakers unmasked might be small, a handful of court orders might be sufficient to deter the most egregious abuse of the social media. Cass Sunstein (2009) has defended this "chilling effect" on the grounds that it "can be an excellent safeguard. Without such an effect," he continues, "the marketplace of ideas will lead many people to spread and to accept damaging falsehoods about both individuals and institutions" (p. 11). The danger, of course, is that the same fear of exposure that moderates the most egregious speech on forums like JuicyCampus might also moderate important speech on matters of public concern. The trick is achieving a Goldilocks-like balance—not too much, not too little, but just enough accountability to produce the desired result. In this context, that would mean unmasking enough speakers to deter outrageous claims and, at the same time, protecting anonymity enough so that speakers feel secure in sharing important information. Achieving such a delicate balance, unfortunately, is all but impossible.

Over the longer term, some have questioned whether forums featuring anonymous communication are sustainable. "Any medium with neither access cost nor legal intervention will generate a high noise-to-signal ratio," Levmore (2010) argues, "and most communications will be of low value. Thus, a free wall for graffitists, without periodic cleansing, will be of low value to readers and juvenile authors alike, because the volume of junk will reduce the audience" (p. 57). This explains why sites like JuicyCampus, "where there is no active censor," will quickly "sink to the lowest common denominator" (p. 58). Given enough time, however, participants will self-select and most will eventually lose interest in such base and vile content.

To prevent this from happening, a measure of accountability is required, and that is why many newspaper sites now require users to register before they post comments. This policy holds speakers nominally accountable, and that is generally sufficient to prevent the most egregious comments. The resulting posts might be more temperate, but the content appeals to a broader audience. As an alternative to registration, other sites have developed formal policies that allow them to exercise some control over content. For example, Amazon.com polices and removes objectionable reviews from its web site. This is expensive, but it also reduces the noise problem. The Internet does not have an anonymity problem, Levmore (2010) argues: "the real problem is one of offensive or noise information, and the challenge is to devise low-cost methods of improving the exchange of information" (p. 62).

The experience of Wikipedia may be instructive on this point. Based on a common-collaborative model, Wikipedia allows anyone to edit articles and it also permits users to edit anonymously. This creates a real possibility for mischief, especially with respect for biographies of living persons, as is illustrated by the Seigenthaler incident. To prevent this sort of vandalism, Wikipedia developed mechanisms to verify the accuracy of content and to prevent malicious editing. For example, there are standards limiting who can initiate a biography page. Changes to these pages must be approved by an "experienced volunteer editor" before they are posted. Since these pages have been repeatedly targeted, Wikipedia also conducts periodic reviews on the biographies of living persons (Chen, 2010, pp. 282–284).

This sort of self-regulation is hardly a perfect solution. While "more Internet entrepreneurs will limit participation or require identification," Levmore (2010) concludes, "there will remain a few sites that remind us of JuicyCampus, occupying a niche like that in which we find porn shops in city centers, at the periphery of most social interaction" (p. 67). While sensitive souls could avoid these adult establishments, this analogy fails, as search engines do not discount speech from unreliable sources. Google cannot distinguish between a legitimate complaint and a vicious character attack. Even if self-regulation ultimately eliminates the worst offenders, it will come with a price. Speech on sites that require registration or identification will be restrained. Knowing they can be held accountable for their speech, speakers will practice a measure of self-censorship. Such a chilling effect will result in less defamation, but it will also mean there is less truthful speech too. Much of what is lost might be inconsequential, but some legitimate content will be lost for fear of the consequences.

## The Price of Free Speech

This brings the analysis back to the original three questions. With respect to the first question, whether the standards developed for the traditional media extend to the social media, the answer is clear. Traditional defamation law extends to the social media, and this means it will be difficult for aggrieved parties to collect damages for defamatory speech communicated through the social media. Since *New York Times v. Sullivan* (1964), it has been all but impossible for public figures to prevail in defamation actions. This does not mean, however, that redress is impossible, as the outcome of *Simorangkir v. Love* demonstrates. Love's outrageous accusations were unfounded, and the financial settlement suggests Simorangkir might well have prevailed had the case gone to trial. It would, moreover, be easier for a private person to recover damages for defamatory speech disseminated using the social media, as there is a lower standard of fault.

The second question asked whether ISPs or content facilitators could be held responsible for third-party expression. On this point, the case law is also clear. Courts have broadly interpreted Section 230 to confer sweeping immunity on sites like JuicyCampus. This protection is unique, as immunity has not been extended to traditional publishers or broadcasters. Critics of Section 230 have complained that it affords too much protection. To redress these problems, a variety of schemes, including notice-based liability, have been proposed. Imposing such liability, however, comes at a price. Efforts to reign in defamatory content would necessarily chill some legitimate criticism and commentary.

This leads to the third question, which asked whether judges would unmask anonymous speakers, so that aggrieved parties could file a defamation suit against the speaker. While the standards for assessing such motions remain a work in progress, sympathetic judges have already unmasked some anonymous speakers. It is too early to tell, however, whether judges will be able to achieve the balance—enough anonymity to support a robust social media, but not too much anonymity as that would leave aggrieved parties without legal redress against scurrilous rumors and salacious gossip.

By narrowly focusing on these three questions, however, it might be argued that this analysis has missed the larger issue. Instead of focusing on the applicable standards, one might productively question how much constitutional protection should be extended to defamatory speech on the social media. In *New York Times v. Sullivan* (1964), the Supreme Court famously extended First Amendment protection to a class of defamation—that which is directed at officials of the government. Justice William Brennan wrote the following:

> The constitutional guarantees require a federal rule that prohibits a public official from recovering damages for a defamatory falsehood relating to his official conduct unless he proves that the statement was made with "actual malice"—that is, with knowledge that it was false or with reckless disregard of whether it was false or not. (pp. 280–281).

In this instance, however, the speech dealt with the conduct of an elected public official and addressed a matter of public concern. Should the same deference be extended to scurrilous accusations on Wikipedia, outrageous accusations of sexual impropriety on a bulletin board, or a vicious parody of a school administrator on Myspace?

## References

Ardia, D. S. (2010). Free speech savior or shield for scoundrels: An empirical study of intermediary immunity under Section 230 of the Communications Decency Act. *Loyola of Los Angeles Law Review, 43*, 373–506.

Belloni, M. (2011, March 2). Courtney Love to pay $430,000 to settle Twitter defamation case. *Hollywood Reporter.* Retrieved from http://www.hollywoodreporter.com/blogs/thr-esq/courtney-love-pay-430000-settle-163919

Chen, S. (2010). Wikipedia: A republic of science democratized. *Albany Law Journal of Science & Technology, 20*, 247–325.

Citron, D. K. (2010). Mainstreaming privacy torts. *California Law Review, 98*, 1805–1852.

*Cohen v. California*, 403 U.S. 15 (1971).

*Cohen v. Google*, 87 N.Y.S.2d 424 (N.Y. Sup. Ct. 2009).

Communications Decency Act, 47 U.S.C. §230 (1996).

Complaint, *Horizon Group Management LLC v. Amanda Bonnen*. (2009, July 20). Circuit Court of Cook County, Illinois. Retrieved from http://www.citmedialaw.org/sites/citmedialaw.org/files/2009–07–27-Horizon%20Complaint.pdf

Complaint, *Simorangkir v. Love* (2009, March 26). Superior Court of the State of California for the County of Los Angeles. Retrieved from http://www.citmedialaw.org/sites/citmedialaw.org/files/2009–03–26-Simorangkir%20Complaint.pdf

*Cubby v. CompuServe*, 776 F.Supp. 135 (S.D. N.Y. 1991).

Denby, D. (2009). *Snark: A polemic in seven fits.* New York: Simon & Schuster.

*Dendrite International v. John Doe*, No. 3, 342 N.J. Super. 134 (N.J. Super. 2001).

*Doe I and Doe II v. Individuals, whose true names are unknown*, 561 F. Supp. 2d 249 (D. Conn. 2008).

*Doe v. Cahill*, 884 A.2d 451 (Del. 2005).

Dowd, M. (2009, August 25). Stung by the perfect sting. *New York Times*, p. A23.

Electronic Frontier Foundation. (n.d.). Takedown hall of fame. Retrieved from https://www.eff.org/takedowns

Grimmelmann, J. (2008/2009). The Google dilemma. *New York Law School Law Review, 53*, 939–952.

Hafner, K. (2001). *The well: A story of love, death, and real life in the seminal online community.* New York: Carroll & Graf.

*J.S. v. Blue Mountain School District*, 2011 U.S. App. LEXIS 11947 (3rd Cir. 2011).

Lau, T. J. (2011). Government censorship of media: Towards zero net presence. *Notre Dame Journal of Law, Ethics & Public Policy, 25*, 237–277.

Leiter, B. (2010). Cleaning cyber-cesspools: Google and free speech. In S. Levmore & M. C. Nussbaum (Eds.), *The offensive internet* (pp. 177–173). Cambridge, MA: Harvard University Press.

Levmore, S. (2010). The internet's anonymity problem. In S. Levmore & M. C. Nussbaum (Eds.), *The offensive internet* (pp. 50–67). Cambridge, MA: Harvard University Press.

Masnick, M. (2005, January 5). Since when is it illegal to just mention a trademark online *Techdirt*. Retrieved from http://www.techdirt.com/articles/20050105/0132239.shtml

Mayer-Schönberger, V. (2009). *Delete: The virtue of forgetting in the digital age.* Princeton, NJ: Princeton University Press.

McCarthy, A. (2011, March 4.) Courtney Love settles $430,000 Twitter defamation lawsuit. *Huff Post Entertainment*. Retrieved from http://www.huffingtonpost.com/2011/03/04/courtney-love-settles-430_n_831302.html

*McIntyre v. Ohio Elections Commission*, 514 U.S. 334 (1995).

Morgan, R. (2008, March 18). Juicy campus: College gossip leaves the bathroom wall and goes online. *New York Times*. Retrieved from http://www.nytimes.com/2008/03/18/travel/18iht-gossip.1.11208865.html?pagewanted=all

*New York Times v. Sullivan*, 376 U.S. 254 (1964).

Official JuicyCampus Blog. (2009, February 4). A juicy shutdown. Retrieved from http://juicy-campus.blogspot.com/

Order, *Horizon Group Management LLC v. Amanda Bonnen*. (2010, January 27). In the Circuit Court of Cook County, Illinois. Retrieved from http://www.balough.com/uploadedFiles/Dismissal%20Order.pdf

Pasquale, F. (2010). Reputation regulation: Disclosure and the challenge of clandestinely commensurating computing. In S. Levmore & M. C. Nussbaum (Eds.), *The offensive internet* (pp. 107–23). Cambridge, MA: Harvard University Press.

*Reno v. ACLU*, 521 U.S. 844 (1997).

Seigenthaler, J. (2005, November 29). A false Wikipedia "biography." *USA Today*, p. 11A. Retrieved from http://www.usatoday.com/news/opinion/editorials/2005-11-29-wikipedia-edit_x.htm

Seltzer, W. (2010). Free speech unmoored in copyright's safe harbor: Chilling effects of the DMCA on the First Amendment. *Harvard Journal of Law & Technology*, 24, 171–232.

Solove, D. (2010). Speech, privacy, and reputation on the internet. In S. Levmore & M. C. Nussbaum (Eds.), *The offensive internet* (pp. 15–30). Cambridge, MA: Harvard University Press.

Solove, D. J. (2007). *The future of reputation: Gossip, rumor, and privacy on the Internet.* New Haven, CT: Yale University Press.

*Stratton Oakmont v. Prodigy*, 1995 N.Y. Misc. LEXIS 229 (N.Y. Sup. Ct. 1995).

Sunstein, C. (2009). *On rumors: How falsehoods spread, why we believe them, what can be done.* New York: Farrar, Straus & Giroux.

Wright, C. M. (2009, July 28). Amanda Bonnen, apartment renter, sued for "defamatory" Twitter post about mold. *Huffington Post*. Retrieved from http://www.huffingtonpost.com/2009/07/27/amanda-bonnen-apartment-r_n_245944.html

*Zeran v. America Online*, 958 F. Supp. 1124 (E.D. Va. 1997), aff'd, 129 F.3d 327 (4th Cir. 1998), cert. denied, 524 U.S. 937 (1998).

# · 4 ·

# Cyberharassment and Cyberbullying

## "There Ought To Be a Law"

JULIET DEE

2003: 13-year-old Ryan Halligan receives vulgar instant messages from classmates who accuse him of being gay. After an eighth-grade girl fakes an online relationship with him and then "breaks up" with him at school while other students laugh, Ryan hangs himself (Garlic, 2010).

2005: 13-year-old Jeff Johnston becomes a victim of cyberbullying in Florida. He hangs himself in his closet (Fantz, 2006).

2006: 13-year-old Megan Meier reads a message from "Josh Evans," saying, "The world would be a better place without you." Not realizing that Josh Evans does not exist but is a fabrication of Meier's adult neighbor, Lori Drew, Meier hangs herself in her bedroom (Steinhauer, 2008).

2009: In Cincinnati, Ohio, Jesse Logan sends a nude picture of herself as a text message on her cell phone to her boyfriend. After her boyfriend forwards the photo to their classmates at school, her classmates harass her to the point where she finally commits suicide.

2010:    Six high school students gang up on 15-year-old Irish immigrant Phoebe Prince, vilifying her on Facebook and sending her text messages calling her a slut and a whore and saying that she deserves to die. Prince walks home after school and hangs herself (Eckholm & Zezima, 2010).

2010:    Rutgers University students Dharun Ravi and Molly Wei set up a hidden webcam and stream a live Internet feed of an intimate encounter between freshman Tyler Clementi and another man. Three days later Clementi throws himself off the George Washington Bridge (Foderaro & Hu, 2010).

2010:    Police in Cleveland, Texas, arrest 18 teenage boys and young men for the gang rape of an 11-year-old girl. Some of the rapists have recorded the assaults on their cell phones and passed the videos of the gang rape around at school (McKinley, 2011).

Ryan Halligan, Jeff Johnston, Megan Meier, Jesse Logan, Phoebe Prince, and Tyler Clementi were just a few of the thousands of teenagers and young adults who have been victims of cyberbullying. Of course, most young victims do not commit suicide, but cyberbullying has become so rampant that courts, state legislatures, and Congress have struggled to formulate either judicial or legislative responses to this problem.

This discussion will focus on both judiciary and legislative attempts to arrive at either court decisions or statutes designed to deter both cyberharassment and cyberbullying without violating the First Amendment right of free speech. The discussion begins with a look at court decisions involving cyberbullying cases, followed by a look at bills introduced in Congress and also in state legislatures in this ever-developing area of social networking law.

# Definitions of
# Cyberharassment and Cyberbullying

Some legal scholars have begun to use the term *cyberharassment* to describe cases in which students post derogatory comments about their teachers and principals (Turbert, 2009, note 48); this is to distinguish cyberharassment from *cyberbullying*, which describes cases where the victim is the same age as the bully.

Cyberbullying has been defined as

any type of harassment or bullying (teasing, telling lies, making fun of someone, making rude or mean comments, spreading rumors, or making threatening or aggressive comments) that occurs through e-mail, a chat room, instant messaging, a web site (including blogs), text messaging, videos or pictures posted on web sites or sent through cell phones.(Calvert, 2010, pp. 9–10).

## The Quartet of Student Speech Cases: Tinker, Fraser, Hazelwood, and Morse

Although the U.S. Supreme Court has never ruled specifically on a case involving cyberbullying, and it has not ruled on the extent to which a school district can discipline students for off-campus speech, the high court has handed down four decisions on the degree of First Amendment protections for student speech. Lower courts have relied on these four decisions in deciding the cases involving cyberbullying. The four decisions are *Tinker v. Des Moines Independent Community School District* (1969),[1]*Bethel School District No. 403 v. Fraser* (1986),[2]*Hazelwood School District v. Kuhlmeier* (1988),[3] and *Morse v. Frederick* (2007).[4]

Viewed together, this "quartet" of *Tinker, Fraser, Hazelwood,* and *Morse* suggests that high school administrators can exert control over disruptive speech if it is lewd, offensive, or perceived to promote illegal activity. What is far less clear is the extent to which school administrators have jurisdiction over students' off-campus speech in e-mail messages, blogs, text messages, and social networking sites such as Myspace and Facebook.

## Codes of Student Conduct on College Campuses

At some universities, administrators have begun to revise their codes of student conduct to include students' online behavior when the students are off-campus. For example, in 2009, University of Wisconsin Dean Lori Berquam notified all students that they could be charged with misconduct for off-campus (as well as on-campus) behavior. This includes online behavior such as "breach of computer security, invasion of privacy or unauthorized access to university computing resources" (University of Wisconsin System, 2009, 17.06–10–d). The Ohio State University likewise revised its Code of Student Conduct in 2010 to cover off-campus behavior; the code prohibits photographing or videotaping anyone where there is a reasonable expectation of privacy without the person's prior knowledge. This includes taking video or photographic images in showers or locker rooms, dormitories, and bathrooms. Storing, sharing, or distributing such images is also prohibited (Ohio State University Code of Student Conduct, 2010, 3335–23–04(Q).

The University of Delaware likewise has the Policy for Responsible Computer regarding use of the school's computers and servers. In 2007, the university took disciplinary action against chemical engineering student Maciej Murakowski after Murakowski used a university server to post "Maciej's Official Guide to Sex," in which he included numerous graphic descriptions of how to gang-rape and murder women; for example, he wrote, "This is my glove choking a bitch. That bitch has been thoroughly choked" (*Murakowski v. University of Delaware*, 2008, Memorandum Opinion, p. 9). Between 2005 and 2007, several women students and at least one of their parents contacted University of Delaware administrators to express concern about Murakowski's "bizarre behavior"; the young women who lived in Murakowski's residence hall were frightened and disturbed when they saw his web site, and they felt that Murakowski had threatened their personal safety (*Murakowski v. University of Delaware*, 2008, Memorandum Opinion, p. 10). The University of Delaware expelled Murakowski, at which point he filed suit against the university for violating his First Amendment rights.

U.S. Magistrate Judge Mary Pat Thynge held that although Murakowski's postings were vulgar and sophomoric, they did not rise to the level of a "true threat" and did not target any individual student on the campus. She thus ruled that the University of Delaware had violated Murakowski's First Amendment rights and ordered the university to pay him $10 in damages (*Murakowski v. University of Delaware* Judgment, 2008, Order, p. 2). Although Murakowski was not guilty of cyberbullying per se, the university's actions reflect a growing concern around the country regarding the problem of students using university-owned computers and servers to post offensive material.

## Cases Involving Cyberharassment

The distinction between cyberharassment and cyberbullying cases is helpful, due to the fact that there are many more court decisions involving cyberharassment than cyberbullying. This is probably because teachers, principals, and school districts are more likely to resort to legal remedies when students denigrate teachers and school administrators, whereas junior high and high school students are often reticent about reporting that they have become victims of cyberbullying in the first place. Thus, we begin with an examination of cyberharassment cases, first considering cases in which teachers and school administrators have prevailed, followed by cases in which students have won lawsuits against school districts on First Amendment grounds. Because judges also look to precedent, the cases in both categories are considered in chronological order.

## Cyberharassment Cases in Which School Administrators Prevailed

In several cases during the past decade, school administrators have either suspended or expelled students who created web sites that denigrated—and sometimes even threatened—their teachers and principals. Courts have ruled in favor of the school districts and school administrators in the cases that follow on the grounds that the web sites or videos the students created had "materially disrupted" discipline at school, even if the students had created the videos or web sites at home.

Because many of these cases have been discussed elsewhere, the cyberharassment cases are outlined briefly here:

- ~ *J.S. ex rel H.S. v. Bethlehem Area School District (2002)*: A student who solicited funds for a hit man to kill his math teacher, Ms. Fulmer, with pictures of her severed head dripping blood on his web site, "Teacher Sux," was expelled. The Pennsylvania Supreme Court upheld his expulsion.

- ~ *Hillary Transue (Pilkington, 2009)*: Hillary Transue created a Myspace page mocking her high school's assistant principal. School administrators charged her with harassment in juvenile court and she served 3 months in a juvenile detention center.[5]

- ~ *Requa v. Kent School District No. 415 (2007)*: Gregory Requa edited a video that zoomed in on female teachers' buttocks. After he posted the video on YouTube, he was suspended for 40 days. A federal district court in the state of Washington upheld his suspension.

- ~ *Wisniewski v. Board of Education (2007)*: Aaron Wisniewski created an icon with a pistol firing a bullet at a man's head with the words, "Kill Mr. VanderMolen," his teacher. He was suspended, and the U.S. Court of Appeals for the Second Circuit upheld that suspension.

~ *Doninger v. Niehoff (2008)*: Avery Doninger posted an entry to her
   LiveJournal blog in which she referred to the school administrators as "d——bags" and encouraged others to contact school principal Karissa Niehoff to "piss her off even more" (*Doninger v.
   Niehoff*, 2008, p. 206). The U.S. Court of Appeals for the Second
   Circuit upheld the school's decision to punish Doninger for her web
   site by forbidding her to run for senior class secretary and forbidding her to speak at graduation.

~ *People of the State of Colorado v. Davis Temple Stephenson (2008)*:
   Davis Temple Stephenson, a student at Fort Lewis College in
   Colorado from 1999–2003, created a fake web site, pretending to
   be his female professor. The fake profile said that she enjoyed
   being raped, invited anyone to come to her house and rape her, and
   listed her actual home address. The jury found Stephenson guilty
   of criminal libel and Judge David Dickinson sentenced him to 23
   years in prison.

~ *J.S. v. Blue Mountain School District (2010)*: J.S. posted a fake
   Myspace profile of school principal James McGonigle, characterizing McGonigle as a pedophile and sex addict. J. S. was suspended, and the U.S. Court of Appeals for the Third Circuit
   upheld that suspension.

## Cyberharassment Cases in Which Students Prevailed

In the cases outlined earlier, the courts upheld the decisions of the school
administrators to punish the students for their web sites or videos posted
online. In the following cases, students prevailed when they sued their school
districts for violating their First Amendment rights. As in the first category of
cases, the cases are discussed in chronological order.

~ *Emmett v. Kent School District No. 415 (2000)*: Nick Emmett created "mock obituaries" of two friends in response to an English class
   assignment on obituaries. The school district suspended him for 5
   days, but a federal district court in the state of Washington
   enjoined Emmett's suspension, finding that his mock obituaries did
   not materially disrupt classes at Kentlake High School.

~ *Killion v. Franklin Regional School District (2001)*: Zachariah Paul
   mocked athletic director Robert Bozzuto regarding his appearance

and the size of his genitals. School administrators suspended him, but a federal district court in Pennsylvania held that Paul had not created a "substantial disruption" to school activities and ruled in the student's favor.

~ *Mahaffey ex rel. Mahaffey v. Aldrich (2002):* Joshua Mahaffey posted a list titled, "People I wish would die." The school district suspended Mahaffey, but a federal district court in Michigan held that Mahaffey's messages did not constitute "true threats" and ruled in Mahaffey's favor.

~ *I.M.L. v. State of Utah (2002):* I.M.L. described his principal as a "town drunk" and accused the principal of having sex with a high school secretary. Although a juvenile court found I.M.L. guilty of criminal libel, the Utah Supreme Court reversed the ruling on the grounds that Utah's criminal libel statute was unconstitutional.

~ *Neal v. Efurd (2005):* Ryan Kuhl and Justin Neal created a "F—k Greenwood" web site and an online comic portraying Vice Principal Garbo shooting a student in the head. The principal suspended Kuhl and Neal, but a federal district court in Arkansas held that the students' off-campus speech had not substantially disrupted school activities and ruled in favor of them.

~ *Muzzillo v. Davis (2006):* Alex Davis created a fake Myspace profile of his high school science teacher, Robert Muzzillo, suggesting that Muzzillo was gay and wrestled with alligators and midgets. Muzzillo sued Davis for defamation, but a court in Georgia dismissed those charges (Student charged, 2006).

~ *A.B. v. State of Indiana (2008):* A.B. posted a message, "Die...Gobert...die!" to Principal Shawn Gobert. The state of Indiana filed a delinquency petition against A.B., but the Indiana Supreme Court held that there was not enough evidence to find the student guilty of harassment.

~ *Draker v. Schreiber (2008):* Benjamin Schreiber and Ryan Todd appropriated Vice Principal Anna Draker's identity, creating a false Myspace page with Draker's name, photograph, and location of Clark High School in Texas. Schreiber and Todd implied that Draker was a lesbian and posted photos of sexual devices. Draker sued the students for defamation, but a Texas court dismissed her suit.

~ *Layshock v. Hermitage School District (2010)*: Justin Layshock created a "parody profile" of Principal Eric Trosch, describing Trosch as "a big fag," "a big steroid freak," and "a big whore" (pp. 252–253). The school district suspended Layshock, but the U.S. Court of Appeals for the Third Circuit held that the school district had violated Layshock's First Amendment rights and ruled in the student's favor.

# Court Decisions Involving Cyberbullying

We begin our discussion of cyberbullying with an examination of "less serious" cases, in which the victims suffered hurt feelings but tried to rely on either school administrators or litigation to end the cyberbullying. As with the cyber-harassment cases discussed earlier, the less serious cases are outlined briefly here. This discussion ends with a look at the most tragic cases, in which the victims of cyberbullying committed suicide.

## Students Accused of Cyberbullying Who Prevailed in Court

In several cases in which school districts have tried to suspend students accused of cyberbullying, the students have sued the school districts and won their cases on First Amendment grounds, or in one case, managed to get a defamation case dismissed. The cases in which the cyberbullies themselves engaged in litigation and won are outlined briefly here:

~ *Coy ex. Rel. Coy v. Board of Education (2002)*: School administrators expelled Jonathan Coy for creating a "Losers" web site that included photos of three of his classmates. A federal district court judge held that school administrators had violated Coy's First Amendment rights because the web site had not disrupted school discipline.

~ *State of Ohio v. Ellison (2008)*: High school student Ripley Ellison was found not guilty of criminal harassment for posting a photo of classmate Savannah Gerhard, accusing Gerhard of molesting Ellison's younger brother.

~ *J.C. v. Beverly Hills Unified School District (2010)*: A federal district court in California held that the school district was wrong to suspend high school student J C. for posting a YouTube video in which she called her classmate C.C. "spoiled" and a "slut." The court held that the school district could not punish J.C. for off-campus speech (Murphy, 2009).

~ *Finkel v. Facebook* (2009); *Finkel v. Dauber (2010)*: High school student Denise Finkel first sued Facebook for $3 million but the case was dismissed (Epstein, 2009). Then Denise Finkel sued classmates Michael Dauber, Jeffrey Schwartz, Melinda Danowitz, and Leah Herz for defamation after they posted messages on Facebook saying she had contracted AIDS by having sex with horses, baboons, and male prostitutes dressed as firefighters. The Supreme Court of New York in Nassau County held that these insults "did not contain statements of fact, [so] there was no defamation," (p. 697) thus, the court dismissed Finkel's complaint.

## Cyberbullying Cases Settled out of Court or Pending

Although the cyberbullies themselves prevailed in the cases outlined earlier, a Texas father won an out-of-court settlement when three of his daughter's classmates bullied her on Facebook, and the parents of a 14-year-old girl in Georgia filed a defamation suit against their daughter's classmates after they had bullied her on Facebook. These cases are outlined here.

~ *Jason Medley (2011)*: In 2011, Houston attorney Jason Medley learned that his daughter had become the victim of vicious bullying by three classmates who had posted a video about her on Facebook. The video defamed his daughter in terms that were explicit and sexually derogatory. Several months later, Medley reached an out-of-court settlement with the three girls who had defamed his daughter. The settlement required an apology from the girls and a donation to the Center for Safe and Responsible Internet Use (Turner & Bluestein, 2012).

~ *Alex Boston (2012)*: Fourteen-year-old Alex Boston, a student at the Palmer Middle School near Atlanta, Georgia, became the victim of two classmates who created a phony Facebook page with Boston's name and information and a photograph of her that was doctored to make her face appear bloated. The Facebook page implied that Alex smoked marijuana and spoke a made-up language called "Retardish" (Turner & Bluestein, 2012, p. A2). After trying for an entire year to get Facebook to take down the phony profile, Boston and her parents filed a libel suit against her two classmates, at which point Facebook took down the page (Turner & Bluestein, 2012). The lawsuit was filed in May 2012; the case is pending.

## Cyberbullying Cases Ending in Suicide

During the past decade, there have been several high-profile cases involving teenagers Ryan Halligan, Jeffrey Johnston, Jesse Logan, Megan Meier, Phoebe Prince, and Tyler Clementi, who committed suicide after being victimized by vicious cyberbullying. As a result of these suicides, some state legislatures have passed legislation and several bills have been introduced in Congress with the purpose of deterring cyberbullying.

~ *Ryan Halligan (2003):* When Ryan Halligan was in eighth grade, a bully at his school in Vermont spread a rumor via instant messaging that 13-year-old Halligan was gay. Then a popular girl at the school pretended, via instant messaging, that she liked Halligan, but later she shared the details of her online conversations with other students with the purpose of humiliating Halligan. In front of other students at school, the girl told Halligan that she would never want anything to do with such a "loser." In response, Halligan said that "it was girls like her that made him want to kill himself" (Garlic, 2010, p. 32). Soon afterward, Halligan hanged himself. A year later, in May 2004, Ryan's father, John Halligan, persuaded the state legislature to pass Vermont's Bully Prevention Law, which established procedures for school administrators to deal with both traditional bullying and cyberbullying (Garlic, 2010).

~ *Jeffrey Johnston (2005):* Although it is not clear whether Ryan Halligan reported the bullying against him to school administrators, Jeffrey Johnston repeatedly reported that Robert Roemmick had been bullying him for 2 years, and school administrators did nothing. Throughout all of seventh and eighth grade, Roemmick used Internet attacks to bully straight-A student Johnston, calling Johnston "creepy," a "stalker," and worse. Johnston reported the cyberbullying to school officials, who did nothing to stop Roemmick's vicious attacks. After 2 years, when he was a freshman in high school at age 15, Johnston hanged himself in his room. In

2008, his mother, Debbi Johnston, completed her 3-year mission to get the Florida legislature to pass the Jeffrey Johnston Stand Up for All Students Act, which requires schools to adopt policies to discourage both traditional bullying and cyberbullying or risk losing state funding (Chang, Owens, & Brady, 2008).

~ *Jesse Logan (2009):* Whereas the parents of Ryan Halligan and Jeffrey Johnston managed to persuade the state legislatures of Vermont and Florida to pass legislation dealing with cyberbullying, state legislatures are just beginning to consider legislation dealing with *sexting*. Jesse Logan was a student at Sycamore High School in Cincinnati, Ohio. She sent a nude photo of herself in a cell phone text message (called *sexting*) to her boyfriend, and he forwarded the photo to numerous classmates at school. What began as cyberbullying by Logan's boyfriend then turned into traditional bullying; Logan's classmates harassed her to the point that she refused to go to school. After attending the funeral of an acquaintance, Logan went home and hanged herself in her closet (Celezic, 2009).

~ *United States v. Drew (2009):* Unlike the case of Jesse Logan, which received little attention from the news media, the incident involving Megan Meier was one of the first cyberbullying cases to receive national attention. After Meier had a falling out with 13-year-old Sarah Drew, Drew's mother, 47-year-old Lori Drew, created a fictitious 16-year-old boy, "Josh Evans," who contacted Meier via Myspace and said he had a crush on her. Drew's motive in creating "Josh Evans" was to find out what Meier thought of Sarah. (Sarah Drew considered Meier to be her nemesis.) After several weeks of flirting with Meier, "Josh Evans" told Meier he was moving away. He was suddenly hostile, telling Meier that he no longer liked her and added, "The world would be a better place without you." Meier wrote back, "You're the kind of boy a girl would kill herself over." She hanged herself that afternoon in her bedroom (Steinhauer, 2008). Drew lived in suburban St. Louis, Missouri; although people were outraged when they learned of Drew's subterfuge, law enforcement officials in Missouri determined that Drew had broken no laws. But U.S. Attorney Thomas O'Brien decided to prosecute Lori Drew in Los Angeles where Myspace's servers are based. O'Brien charged Drew under the Computer Fraud and Abuse Act (CFAA). One issue in the case was whether

Drew had violated the Myspace Terms of Service agreement when she created a fictitious profile. Myspace uses two types of agreements for its terms of service. Those who simply visit Myspace will see a "browse-wrap" agreement that binds them to the terms of service. A browse-wrap agreement is a license agreement that covers access to materials on a web site. Before users browse the web site, they are expected to accept the browse-wrap agreement. Once Drew decided to become a member of Myspace, she had to check a box showing that she accepted the "click-wrap" agreement. A click-wrap agreement or license requires users to communicate their assent by clicking on an "okay" or "I agree" button on a dialog box. The Myspace Terms of Service, which prohibited harassment and providing false information, were not on Myspace's registration page, so a person could become a member without ever viewing the Terms of Service, simply by clicking the "check box" and then clicking the "sign-up" button. A visitor to Myspace had to click on several hyperlinks before finding the Terms of Service, allowing Drew to click the sign-up box without reading those terms. With the "check-the-box" click-wrap agreement on the sign-up page and not specifically on the Terms of Service page, the question of whether Drew had read the Terms of Service was unanswered.

U.S. Attorney O'Brien argued that Drew had violated Myspace's Terms of Service by creating a fictitious account. A Los Angeles jury found Drew guilty of accessing a computer without authorization, meaning that Drew had violated the CFAA. But federal District Court Judge George Wu threw out Drew's conviction on the grounds that the CFAA was not only void-for-vagueness but also that the CFAA was not really applicable to the facts of this case. Judge Wu explained that Congress had passed the CFAA in order to punish hacking; the language of the CFAA forbids a person from accessing a computer without authorization (Computer Fraud and Abuse Act, 1986). But Drew had not hacked into Myspace. Legal scholar Brandon Darden (2010) explains that, "Lori Drew was an insider who had the proper authority to access Myspace's web site, especially since it was not required for her even to read the [Myspace Terms of Service] before she signed up. Once she crossed this threshold, the CFAA was powerless against her subsequent actions because Congress did not originally design the statute for that purpose" (p. 359).

Furthermore, Judge Wu explained, someone who breaches a contract (such as the Terms of Service) might expect civil penalties but would not expect a criminal penalty. Judge Wu declined to find that the CFAA would criminalize every breach of contract such as the Myspace Terms of Service; this would make the CFAA overbroad because it would not provide criteria regarding which of the breaches should merit criminal prosecution (*United States v. Drew*, 2009). In other words, Judge Wu held that Drew's conviction was void for vagueness because he was not convinced that every breach of a social network's Terms of Service is equivalent to accessing the web site "without authorization," which the CFAA would have required (*United States v. Drew*, 2009, p. 467). Despite the fact that Drew created a situation resulting in Meier's suicide, Drew suffered no consequences for her actions because there were no applicable laws regarding cyberbullying in 2006 (Cathcart, 2009, p. A12).

~  *Phoebe Prince (2010)*: Just as the Missouri legislature amended its harassment statute in response to the Megan Meier case, the Massachusetts legislature passed an antibullying law following the suicide of Phoebe Prince. Fifteen-year-old Phoebe Prince moved from Ireland to Massachusetts with her family just before school started in September 2010. She enrolled in South Hadley High School, but within a few months at least six students began bullying her, both in person and through text messages and Facebook. They called her an "Irish slut" because she had briefly dated football player Sean Mulveyhill, which enraged Mulveyhill's former girlfriend Kayla Narey. Narey and her friends sent Phoebe text messages calling her a "druggie," a "whore," and told her she deserved to die (Eckholm & Zezima, 2010, p. A3). Although Phoebe's mother, Anne O'Brien, had complained to school officials at least twice, the bullying continued. In January 2010, Phoebe Prince walked home from school and hanged herself. The taunting and abuse continued even after her death; certain messages had to be removed from her Facebook memorial page. District Attorney Elizabeth Scheibel charged Kayla Narey, Ashley Longe, Sharon Velazquez, Flannery Mullins, Sean Mulveyhill, and Austin Renaud with criminal harassment, violation of civil rights, and stalking. Scheibel also charged Mulveyhill and Renaud with statutory rape. After Prince's suicide, the Massachusetts legislature passed an

antibullying law which requires parents to be notified of a bullying incident.

In May 2011, five of the six defendants, Narey, Longe, Velazquez,
Mullins, and Mulveyhill, pled guilty to criminal harassment in
exchange for 1year of probation and 100 hours of community service, thus avoiding any jail time (Khadaroo, 2011). Prosecutors
dropped the statutory rape charges against the sixth defendant,
Austin Renaud (Moran, 2011).

In December 2011, South Hadley High School settled a lawsuit
that Prince's parents, Jeremy Prince and Anne O'Brien had filed
against school administrators for negligence in failing to do anything to stop the six students from bullying their daughter, despite
the fact that Phoebe Prince had frequently reported the bullying
to a number of teachers. The school district paid Jeremy Prince and
Anne O'Brien $170,000 to settle their claim against the high
school (Doyle, 2011).

~    *Tyler Clementi (2010)*: Just as District Attorney Elizabeth Scheibel
filed criminal charges against the six students who had bullied
Phoebe Prince, New Jersey prosecutors filed criminal charges
against a former Rutgers student who used a hidden webcam to
invade his roommate's privacy. Freshman Tyler Clementi was a tal-
ented violinist who played in the Rutgers University Orchestra.
But after being in college for only a few weeks, Clementi's roommate Dharun Ravi set up a hidden web cam to catch Clementi in
an intimate encounter with another man in his dorm room. Ravi
streamed it live and alerted his friends to watch by sending them
messages on Twitter. When Clementi discovered that Ravi and
Ravi's friend Molly Wei had spied on him from another room as he
kissed a male friend, Clementi asked for advice on the gay chat site
JustUsBoys.com. After agonizing about what to do, Clementi
reported Ravi's actions to his resident advisor. But before university officials had a chance to intervene, Clementi threw himself off
the George Washington Bridge. Prosecutors in New Jersey charged
Ravi with invasion of privacy, bias intimidation (meaning hate
speech), and destruction of evidence (Schweber, 2012).

Commenting on the case, former Federal Prosecutor Henry
Klingeman explained: "This is an area of law that has yet to catch
up with changes in society with the dawn of the social-networking
era. Other than invasion of privacy, there is not a lot a prosecutor

can do with that" (Sherman, 2010, p. 8). Law professors Susan Brenner and Megan Rehberg explained that the federal system makes it a crime to violate someone's privacy by, say, intercepting his telephone calls, but it does not make the dissemination of information about another's private life a crime (Brenner & Rehberg, 2009).

In March 2012, Dharun Ravi went on trial in Middlesex County Superior Court. Prosecutors charged that Ravi "used his webcam and Twitter messages to bully [Tyler Clementi] because of [Clementi's] homosexuality" (Schweber, 2012, p. A21). A jury found Dharun Ravi guilty of invading Clementi's privacy and of bias intimidation, a hate crime. Although Judge Glenn Bermann could have sentenced Ravi to 10 years in prison and could have then deported Ravi to his native India ("Ex-Rutgers Student," 2012), the judge sentenced Ravi to only 30 days in jail and a $10,000 fine to be paid to a fund for victims of hate crimes (Zernike, 2012b). Ravi served only 20 days of his 30-day sentence in the Middlesex County Jail in New Jersey, after which he was released on good behavior (Zernike, 2012a).

# Suggestions from Legal Scholars Regarding Laws Applying to Cyberbullying

Because cyberbullying is a problem that has existed for only a decade, legal scholars are grappling with the question of which laws might be applicable to cyberbullying, not to mention the question of how legislators might draft bills designed to deter cyberbullying without violating the First Amendment. For example, legal scholar Jessica Moy argues that the following:

> Cyberbullying should not be afforded First Amendment protection....These bullying tactics serve only to silence, intimidate and inflict pain on other students and do not encourage any reasoned deliberation....Moreover, cyberbullying silences the victims, denying them any benefit of speech....It is difficult to argue that [shutting down] web sites allowing students to vote on "Who's the biggest slut in school" will substantially diminish the marketplace of ideas. (Moy, 2010, p. 586)

Legal scholar Clay Calvert suggests that courts should resuscitate the fighting words doctrine to punish cyberbullies (Calvert, 2010). At the same time, Calvert reminds us that school administrators can mete out harsh punishments without fear of legal retribution because they have qualified immunity from liability for civil damages. Calvert notes the following:

Public school officials can squelch off-campus student speech posted on the Internet
and get away with it, at least without fear of paying monetary damages, because the
extent of First Amendment protection for such expression simply is not clearly established by the courts. (Calvert, 2009, pp. 87–89)

Legal scholar Leah Ward analyzes Arkansas' Cyberbullying Act of 2007
(Act 115), which allows public school administrators to punish students for
statements they make off campus. Ward predicts that the law will face a constitutional challenge because it does not define the term *substantial disruption*
[of discipline in a school] clearly enough (Ward, 2009, p. 806).

Legal scholar Michael Gordon (2009) notes that the North Carolina
General Assembly, hoping to prevent another tragedy like that of Megan
Meier, passed a law entitled, "Protect Our Kids/Cyberbullying Misdemeanor,"
which makes it a crime to "build a fake profile or web site, pose as a minor in
an Internet chat room, an electronic mail message or an instant message, or follow a minor online or into an Internet chat room…" (pp. 55–56).

Legal scholar Kevin Turbert notes that in 2006, New Jersey assemblymen
(apparently unaware that Section 230 of the Communications Decency Act
[CDA; 1996] provides Internet Service Providers [ISPs] with immunity from
liability in defamation cases) introduced a bill that would have made ISPs liable
for false or defamatory messages posted on public forum web sites. When the
assemblymen realized that Section 230 of the CDA provides a "safe harbor" for
ISPs from any cause of action due to defamatory or threatening speech, they
withdrew the bill (Turbert, 2009, p. 679). Legal scholar Alison King suggests
that "Congress should revise [Section 230] of the CDA to create notice-based
liability for ISPs and web site operators, thereby requiring them to remove
defamatory content upon notification" (King, 2010, p. 878).

Legal scholar Allison Hayes reminds us that the courts provide an adequate
remedy if someone's behavior constitutes a "true threat" (Hayes, 2010, p. 287).
Law professors Susan Brenner and Megan Rehberg analyze the applicability of
laws against stalking, harassment, defamation, invasion of privacy, making
true threats, and identity theft, all of which prosecutors have tried to use to pursue cyberbullies (Brenner & Rehberg, 2009). Like Brenner and Rehberg, legal
scholar Darryn Beckstrom also examines the "true threat" doctrine and its
applicability to cyberbullying (Beckstrom, 2008, p. 306).(In *Watts v. United
States* [1969)] the U.S. Supreme Court held that the First Amendment does not
protect "true threats.")

# Legislative Efforts: State Laws and School Districts' Policies Targeting Cyberbullying

At present, 22 states have passed laws to combat cyberbullying.[6] Legal scholar Colleen Barnett (2009) has created two tables indicating the types of harm the state cyberbullying statutes seek to address, with categories such as "substantial disruption" and "creation of intimidating, threatening, abusive and/or hostile education environment," for example (pp. 623–624). In general, state statutes direct school districts to establish policies to deter cyberbullying, providing for students to be suspended or expelled. The Illinois cyberbullying law is an exception; it stands alone in criminalizing cyberbullying beyond the bounds of the public schools' jurisdiction (King, 2010). In addition to cyberbullying laws already in effect, legislators in at least six states, Hawaii, Kentucky, Maine, Massachusetts, New York, and Delaware, have introduced cyberbullying bills.

# Bills Introduced in Congress to Deter Cyberbullying

As the problem of cyberbullying has grown in magnitude, several member of Congress have introduced bills that would deter or punish cyberbullying. Senator Bob Casey (D-PA) introduced the Safe Schools Improvement Act of 2010, which is geared toward preventing traditional bullying rather than cyberbullying (Wallace, 2011). Representative Linda Sanchez (D-CA) introduced the Megan Meier Cyberbullying Prevention Act (2009), which would amend Title 18, Chapter 41, of the U.S. Code as follows:

> Whoever transmits in interstate or foreign commerce any communication, with the intent to coerce, intimidate, harass, or cause substantial emotional distress to a person, using electronic means to support severe, repeated, and hostile behavior, shall be fined under this title or imprisoned not more than two years, or both.

Senator Frank Lautenberg (D-NJ) and Representative Rush Holt (D-NJ) introduced the Tyler Clementi Higher Education Anti-Harassment Act of 2011. This bill would require all colleges that receive federal aid to amend their harassment policies and to recognize cyberbullying as a threat to college students. The bill would also provide grants to help colleges establish anti-bullying programs on campus (Heyboer, 2010, p. 31).

# Attempts to Discourage Cyberbullying in Asia and Europe

The United States is not alone in facing the problem of cyberbullying. In South Korea, a popular singer named Yuni faced a barrage of online accusations that she had gotten cosmetic surgery in 2007. She hanged herself (Lee, 2008). Several months later, one of South Korea's hottest movie actresses, Choi Jin Sil, became the victim of a false, vicious rumor accusing her of being a loan shark and linking her to the suicide of an actor who had run up more than $2 million in debts. This rumor was utterly false, but after "hundreds of thousands of chat-room users…added their own attacks on Choi's morals and character" (Lee, 2008, p. 51) she locked herself in a bathroom and hanged herself. In response, legislators in South Korea's Grand National Party proposed a new law that would allow prosecutors to press charges for online libel (Lee, 2008).

In an attempt to prevent suicides resulting from cyberbullying, such as those of Yuni and Choi in South Korea, the European Union developed a Safer Internet Programme in 1999. One of its programs is called InSafe, which promotes "safe, responsible use of the Internet and mobile devices" (InSafe, 2012). The Safer Internet Programme also sponsors the International Association of Internet Hotlines (INHOPE), an organization that warns parents and children about pedophiles and predators using social networks. The European Network and Information Security Agency (ENISA) released a 2011 report urging law enforcement agencies in all European countries to acquire "additional knowledge and resources" in order to properly cover regulatory issues[7] (p. 7). The report creates a hypothetical profile of a teenage girl named Kristie and then advises parents on issues such as cyberbullying and the dangers from pedophiles who can easily target individual child victims through social networks (ENISA, "Cyber-bullying and online grooming").

# Conclusion

As has been the case with issues such as anonymous defamation and invasion of privacy on the Internet, once again the legal system is struggling to catch up with runaway problems such as cyberharassment and cyberbullying. It is worth noting that, with regard to the cyberharassment cases considered here, students who filed suit against school districts for violating their First Amendment rights prevailed in court more frequently than did the school districts accusing the students of cyberharassment of teachers and principals.

Teachers who attempted to rely on defamation suits against students for cyberharassment nearly always lost; for example, in *I.M.L. v. Utah, Muzzillo v. Davis,* and *Draker v. Schreiber,* judges dismissed these civil libel cases against the offending students. The only exception was the Colorado case involving Davis Temple Stephenson, in which a jury found Stephenson guilty of *criminal* libel for creating a fake web page pretending to be his college professor, inviting men to come to her home and rape her and giving her correct home address. In this case, the jury no doubt concluded that Stephenson had created a "true threat" to his professor's personal safety and the judge sentenced him to 23 years in prison (*People of the State of Colorado v. Davis Temple Stephenson,* 2008).

In the cyberharassment cases in which the school districts prevailed, judges did not hesitate to uphold suspensions or expulsions of students for their off-campus speech, provided the "speech" was substantially disruptive to school discipline. When students created web sites showing a teacher's severed head (*J.S. v. Bethlehem Area School District*), a person shooting his math teacher in the head (*Wisniewski v. Board of Education*), or a YouTube video zooming in on a female teacher's buttocks and cruelly mocking her (*Requa v. Kent School District*), we might assume that the "speech" in these cases was so egregious and so disruptive that courts backed up the school districts' decisions to suspend or expel the students.

With regard to the cyberbullying cases in which the bullies prevailed in court, as with the cyberharassment cases, the court dismissed the defamation case that Denise Finkel filed. But Jason Medley, the father of a girl who was bullied in Houston, reached an out-of-court settlement with the three cyberbullies who had targeted his daughter. Out-of-court settlements can be viewed as a victory for the plaintiffs, although they do not set a precedent. Furthermore, Alex Boston has filed a defamation suit against two of her middle school classmates in Georgia; since she filed her suit in April 2012, her case is pending. With regard to the other three cases in which the bullies prevailed, the courts held that the bullies' off-campus speech had not substantially disrupted school discipline.

The most disturbing cyberbullying cases are those that resulted in the victim committing suicide, as in the cases of Ryan Halligan, Jeffrey Johnston, Jesse Logan, Megan Meier, Phoebe Prince, and Tyler Clementi. The parents of Ryan Halligan, Jeffrey Johnston, and Jesse Logan did not pursue criminal charges against the bullies. Although she was charged under CFAA, a California court held that the CFAA was not really applicable to Lori Drew's behavior toward Megan Meier; thus, although she instigated Megan Meier's suicide, Drew herself was unscathed. Prosecutors in Massachusetts filed criminal charges against the bullies who drove Phoebe Prince to suicide, and prosecutors in New Jersey won their case against Dharun Ravi when they charged him with violating Tyler Clementi's privacy. Five years ago, cyberbullying was an offense in search of a law. But state legislatures and state and federal courts are moving quickly to write laws and provide a legal framework by which school districts can discourage cyberbullying.

## Notes

1. In *Tinker*, the U.S. Supreme Court held that students could wear black armbands to protest against the Vietnam War, provided that they did not "materially" interfere with school discipline.
2. In *Bethel School District*, the U.S. Supreme Court upheld the right of school administrators to suspend Matthew Fraser for giving a speech with sexual references and double entendres because Fraser's speech was "substantially disruptive" to maintaining order in the high school.
3. In *Hazelwood*, the U.S. Supreme Court upheld the school principal's right to delete articles about teen pregnancies and parents' divorces from the high school newspaper; the principal's motives were to protect the privacy of pregnant students and parents of students undergoing acrimonious divorces.
4. In *Morse*, the U.S. Supreme Court held that school administrators could suspend Joseph Frederick for holding up a "Bong Hits for Jesus" banner during an Olympic Torch parade; Chief Justice John Roberts explained that school administrators may punish students for "promoting illegal drug use" in violation of school regulations.
5. In a strange twist, Pennsylvania authorities eventually discovered that Judge Mark Ciavarella, who had sentenced Hillary Transue, had taken, along with his colleague Michael Conahan, "more than $2.6 million in kickbacks to send teenagers to two privately run juvenile detention centers, run by PA Child Care and Western PA Child Care (Urbina & Hamill, 2009, p. A22). Unfortunately, by the time this judicial graft was discovered, Transue had already served 3 months in PA Child Care.
6. The 22 states with cyberbullying laws are Arkansas (Ark. Code Ann. S 6–18–514, 2007); California (Cal. Educ. Code SS 32260–96, 2002); Delaware (Del. Code Ann. Title 14, S 4112D, 2007); Florida (Fla. Stat. S 1006.147, 2009); Idaho (Idaho Code ann. S 18–917A); Illinois (720 Ill. Comp. Stat. Ann. 5/12–7.5); Indiana (Ind. Code S 20–30–5.5–3); Iowa (Iowa Code S 280.28); Kansas (Kan. Stat. Ann. S 72–8256); Maryland (Md. Code Ann. Educ. S 7–424.1, 2008); Minnesota (Minn. Stat. S 121A.0695, 2008); Missouri (Mo. Ann. Stat. S 565.090, West 1999 & Supp. 2009); Nebraska (Neb. Rev. Stat. S 79–2, 137); New Jersey (N.J. Stat. Ann. S 18A:37–15); North Carolina (Gen. Stat. Chap. 14, Section 14–458.1, 2007); Oklahoma (S.B. 1941, 51st Leg., 2d Sess. Okla. 2008); Pennsylvania (24 Pa. Cons. Stat. Ann. S 13–1303.1-A); Rhode Island (R.I. Gen. Laws S 16–21–26); South Carolina (S.C. Code Ann. S 59–63–120); and Washington (Wash. Rev. Code S 28A.300.285).

7. The report also warns that pedophiles can use certain software such as SocNetMiner and AnalyzeThem to troll through social networks such as Facebook and Myspace for information about individual children's whereabouts.

# References

*A.B. v. State of Indiana*, 863 N.E. 2d 1212 (2007); superseded, 885 N.E. 2d 1223 (Ind. 2008).

Barnett, C. (2009). A new frontier and a new standard: A survey of and proposed changes to state cyberbullying statutes. *Quinnipiac Law Review, 27*, 579.

Beckstrom, D. C. (2008). State legislation mandating school cyberbullying policies and the potential threat to students' free speech rights. *Vermont Law Review, 22*, 283.

*Bethel School District No. 403 v. Fraser*, 478 U.S. 675 (1986).

Brenner, S.W., & Rehberg, M. (2009). "Kiddie crime?" The utility of criminal law in controlling cyberbullying. *First Amendment Law Review, 8*, 1.

Calvert, C. (2009). Qualified immunity and the trials and tribulations of online student speech: A review of cases and controversies from 2009. *First Amendment Law Review, 8*,86.

Calvert, C. (2010). Fighting words in the era of texts, IMs and e-mail: Can a disparaged doctrine be resuscitated to punish cyber-bullies? *DePaul Journal of Art Technology & Intellectual Property Law, 21*,1.

Cathcart, R. (2009, July 3). Conviction is tossed out in Myspace suicide case. *The New York Times*, p. A12

Celezic, M. (2009, March 6). Her teen committed suicide over "sexting." Retrieved fromhttp://www.msnbc.msn.com/id/29546030/

Chang, J., Owens, L., & Brady, J. (2008, May 2). Mom's campaign for Florida anti-bully law finally pays off. ABC News. Retrieved from http://abcnews.go.com/GMA/story?id=4774894&page=1#.UFuDRa5XLIQ (site visited 20 September, 2012).

Communications Decency Act of 1996 (47 U.S.C. Section 230).

Computer Fraud and Abuse Act. (1986). 18 U.S.C.A. S 1030(a)(2)(C).

*Coy ex. Rel. Coy v. Board of Education*, 205 F. Supp. 2d 791 (N.D. Ohio 2002).

Darden, B. (2010). Definitional vagueness in the CFAA: Will cyberbullying cause the Supreme Court to intervene? *Southern Methodist University Science & Technology Law Review, 13*, 329.

*Doninger v. Niehoff*, 514 F. Supp. 2d 199 (D. Conn. 2007), affirmed, 527 F. 3d 41 (2d Cir. 2008).

Doyle, M. (2011, December 29). Phoebe's school settles suicide for $170,000. (Ireland) *The Sun*, p. 12.

*Draker v. Schreiber*, 271 S.W. 3d 318 (Texas Ct. App. 4th District, filed August 13, 2008).

Eckholm, E., & Zezima, K. (2010, April 2). Questions for school after teenager's suicide. *New York Times*, pp. A1, A3.

*Emmett v. Kent School District No. 415*, 92 F. Supp. 2d 1088 (W.D. Wa. 2000).

Epstein, R. J. (2009, March 2). Oceanside teen sues Facebook, ex-classmates for $3 million. Retrieved from http://www.newsday.com/long-island/nassau/oceanside-teen-sues-facebook-ex-classmates-for-3m-1.896009 (site visited 20 September, 2012).

European Network and Information Security Agency (ENISA). (2011, October 31). Cyber-bullying and online grooming: Helping to protect against the risks: A scenario on data mining/profiling of data available on the Internet.

Ex-Rutgers student found guilty in webcam suicide case. (2012, March 17). *News Journal*, p. A3.

Fantz, A. (2006, September 5). Immortalizing a lost child; some South Florida parents find solace in raising money or changing laws to create a legacy in honor of their offspring. *Miami Herald*, p. A1.

*Finkel v. Dauber*, 906 N.Y.S. 2d 697 (Supreme Court of New York, Nassau County, July 22, 2010).

*Finkel v. Facebook*, 2009 N.Y. Misc. LEXIS 3021 (Supreme Court of New York, New York County, filed September 16, 2009).

Foderaro, L. W., & Hu, W. (2010, October 1). Student's online musings point to state of mind before a suicide. *New York Times*, p. A18.

Garlic, T. N. (2010, October 17). A bereft father urges action against bullies; Vermont teen took his own life in 2003. *Star-Ledger*, p. 32.

Gordon, M. R. (2009). The best intentions: A constitutional analysis of North Carolina's new anti-cyberbullying statute. *North Carolina Journal of Law & Technology Online*, *11*, 48.

Hayes, A. E. (2010). From armbands to douchebags: How *Doninger v. Niehoff* shows the Supreme Court needs to address student speech in the cyber age. *Akron Law Review*, *43*, 247.

*Hazelwood School District v. Kuhlmeier*, 484 U.S. 260 (1988).

Heyboer, K. (2010, November 19). Bill targets bullying on college campuses; federal proposal follows Rutgers suicide. *Star-Ledger*, p. 31.

*I.M.L. v. State of Utah*, 61 P. 3d 1038 (Utah 2002).

InSafe. (2012, March 11). Retrieved from http://www.saferinternet.eu/web/insafe-inhope/about-inhope

*J. C. v. Beverly Hills Unified School District*, 711 F. Supp. 2d 1094 (C.D. Cal. 2010).

*J.S. ex rel v. H.S. Bethlehem Area School District*, 807 A. 2d 847 (Pa. 2002).

*J.S. v. Blue Mountain School District*, 593 F. 3d 286 (3rd Cir. 2010).

Khadaroo, S. T. (2011, May 5). Phoebe Prince bullies sentenced, but how do they make things right? *Christian Science Monitor*. Retrieved from http://www.highbeam.com/doc/1G1-255554457.html (site visited 20 September, 2012).

*Killion v. Franklin Regional School District*, 136 F. Supp 2d 446 (W.D. Pa 2001).

King, A. V. (2010, April). Constitutionality of cyberbullying laws: Keeping the online playground safe for both teens and free speech. *Vanderbilt Law Review*, *63*, 845.

*Layshock v. Hermitage School District*, 593 F. 3d 249 (3rd Cir. 2010).

Lee, B. J. (2008, October 27). Death by web posts: South Koreans call it "cyberviolence," and they agree it needs to stop. *Newsweek*, p. 51.

*Mahaffey ex rel. Mahaffey v. Aldrich*, 236 F. Supp. 2d 779 (E.D. Mich 2002).

McKinley, J. (2011, March 9). Vicious assault shakes Texas town. *New York Times*, pp. A13, A16.

Megan Meier Cyberbullying Prevention Act. 2009 H.R. 1966; 111 H.R. 1966.

Moran, B. (2011, May 14). Charge-drop teen thanks Phoebe clan: Rape accused in clear. (Ireland) *The Sun*, 15.

*Morse v. Frederick*, 551 U.S. 393 (2007).

Moy, J. (2010). Beyond "the schoolhouse gates" and into the virtual playground: Moderating student cyberbullying and cyberharassment after *Morse v. Frederick*. *Hastings Constitutional Law Quarterly*, *37*, 565.

*Murakowski v. University of Delaware*, Judgment Order, C.A. No. 07–475-MPT (U.S. District Court for the District of Delaware, filed September 4, 2008).

*Murakowski v. University of Delaware*, Memorandum Opinion, C.A. No. 07–475-MPT (U.S. District Court for the District of Delaware, filed September 4, 2008).

Murphy, P. (2009, December 18). Commentary: Student "cyberbully" wins free speech case. Lawyers USA. Retrieved from http://lawyersusaonline.com/benchmarks/2009/12/18/student-cyberbully-wins-free-speech-case/. Site visited 20 September 2012.

*Neal v. Efurd*, Civ. No. 04–2195 (W.D. Ark. Feb 18, 2005).

Ohio State University Code of Student Conduct. (Revised 2010, May 14). 3335–23–04(Q). Retrieved from http://trustees.osu.edu/rules/code-of-student-conduct.html (site visited 20 September, 2012).

*People of the State of Colorado v. Davis Temple Stephenson*, 2008 Colo. App. LEXIS 1324 (Court of Appeals of Colorado, Division 3, filed August 14, 2008).

Pilkington, E. (2009, March 7). Jailed for a Myspace parody, the student who exposed America's cash for kids scandal: Judges deny kickbacks for imprisoning youths. *The Guardian (London)*, p. 21.

*Requa v. Kent School District No. 415*, 492 F. Supp. 2d 1271 (W.D. Wash. 2007).

Safe Schools Improvement Act of 2010. S.3739; 111 S. 3739.

Schweber, N. (2012, March 8). In video, jury hears student admit posting on roommate. *New York Times*, pp. A21–A22.

Sherman, T. (2010, October 1). Experts: In Rutgers web case, New Jersey law is cloudy. *Star-Ledger*, p. 8.

*State of Ohio v. Ellison*, 900 N.E. 2d 228 (Ohio Ct. App. 2008).

Steinhauer, J. (2008, November 27). Woman guilty in web fraud tied to suicide. *New York Times*, pp. A1, A25.

Student charged for making fun of teacher. (2006, May 16). Retrieved from http://www.wsbtv.com/news/9219518/detail.html

*Tinker v. Des Moines Independent Community School District*, 393 U.S. 503 (1969).

Turbert, K. (2009). Faceless bullies: Legislative and judicial responses to cyberbullying. *Seton Hall Legislative Journal*, 33, 651.

Turner, D., & Bluestein, G. (2012, April 27). Victims of cyberbullying fight back with lawsuits: Georgia case shows limitations of prosecution. *News Journal*, p. A2.

Tyler Clementi Higher Education Anti-Harassment Act of 2011.H.R. 1048; 2011 S. 540.

*United States v. Drew*, 259 F.R.D. 449 (C.D. Cal. 2009).

University of Wisconsin System (UWS) Chapters 17 & 18. (Revised 2009, Sept. 1). Student non-academic disciplinary procedures. Retrieved from http://students.wisc.edu/saja/pdf/New UWS%2017.pdf

Urbina, I., & Hamill, S.D. (2009, Feb. 13). Judges plead guilty in scheme to jail youths for profit. *New York Times*, p. A22.

Wallace, J. A. (2011). Bullycide in American schools: Forging a comprehensive legislative solution. *Indiana Law Journal*, 86, 735.

Ward, L. M. (2009). Suspended on Saturday? The constitutionality of the Cyberbullying Act of 2007. *Arkansas Law Review*, 62, 783.

*Watts v. United States*, 394 U.S. 705 (1969).

*Wisniewski v. Board of Education of the Weedsport Central School District*, 494 F. 3d 34 (2d Cir. 2007), cert. denied, 128 S. Ct. (2008).

Zernike, K. (2012a, June 20). Jail terms ends after 20 days for a former Rutgers student. *New York Times*, p. A26.

Zernike, K. (2012b, May 31). Judge defends penalty in Rutgers spying case, saying it fits crime. *New York Times*, p. A20.

# · 5 ·

# Brains and Behavior

## Addressing Amplified Adolescent Visibility in the Global Village

MARY ANN ALLISON & ERIC ALLISON

Social media connect young people in new and powerful ways. McLuhan and Powers (1992) and Postman (1992) are among the many scholars to point out that any technology has both intended and unintended consequences.

Reports of online harassment and cyberbullying of adolescents, with significant consequences including extended depression, social withdrawal, and suicide, are becoming increasingly common in the press and postings online. A related, often-publicized concern is adolescent exposure to potentially harmful information, images, and strangers on the Internet and lack of care on the part of young people concerning the information they share about themselves. Sexting, for example, is quite different from showing a romantic partner a nude or partially nude physical picture due to the ease with which a digital picture can be sent to others.

Research studies that seek to determine the extent of these activities and the resulting harm yield different results because there are no standard definitions of these activities, and the populations studied (i.e., age, characteristics, and time periods) vary. We include highlights from several studies here.

Hinduja and Patchin (2010) summarized the range of results of seven studies that took place between 2004 and 2010, reporting the following:

- Lifetime cyberbullying victimization rates (bullied at least once in a lifetime) varied from a low of 19% to just over 40% (p. 1).
- Lifetime cyberbullying offending rates ranged from 11.5% to just over 20% (p. 2).

In a 2007 classroom-based study, Hinduja and Patchin found that 42.9% of 2,000 middle school students had experienced at least one upsetting incident online in the past 30 days (Corwin Press, 2008, p. 1).

In addition, the distancing features of mediated communication prompt behavior less likely to occur in face-to-face situations. Studies show that rudeness and mean behavior are increasing and that young people will post or send intimate pictures and information without regard for the amplification features of digital media. For example, Wolak, Mitchell, and Finkelhor (2007, p. 53), found that between 2000 and 2005, the number of young people who admitted to being rude and harassing others online doubled, from 14% in 2000 to 28% in 2005. Approximately 75% of the self-reporting offenders said that these behaviors were directed to people they knew offline (p. 53) and, of these, many said they "used the Internet to harass or embarrass someone they were mad at" (pp. 10–11).

In a 2009 survey (Cox Communications) 76% of teenagers said they were aware of—and at least somewhat concerned about—the risks of posting personal information on the Internet (p. 20), and one in four reported knowing "someone who has had something bad happen to them because of information posted electronically" (p. 8). Even with this knowledge, 72% reported posting personal information on public blogs or social networking sites, 62% have posted photos of themselves, 50% have posted their real age, and more than 40% have posted the name of their school and the city where they live (p. 18). And although sexting—sending nude or nearly nude pictures or videos of one's self—is not as common as many believe, with only 4% of teens reporting sending such images, the fact that it is easy to pass along such images resulted in approximately 15% of teens reporting that they had received sexting images (Lenhart, 2009).

## Overview

In this chapter, we begin by looking at problems of bullying, misbehaving, and risky adolescent experimentation throughout history and then explore the change in the current communication environment that amplifies both this behavior and its consequences. More and stricter laws and law enforcement are among the solutions often proposed to deal with cyberbullying and sexting.

To consider whether this solution is likely to be effective, we turn to the new evidence that shows that the prefrontal lobe of the brain—the part of the brain that deals with consequences—does not fully develop until between the ages of 20 and 25, thus many adolescents are not fully aware of the potential consequences of their online behavior. This has two implications: first, in order to be effective, the U.S. legal system should take the developmental facts into account and, second, laws are not likely to be sufficient to change behavior and thus should be supplemented by other approaches. A brief review indicates that U.S. law does not, in many cases, take young people's incomplete mental development (often referred to in legal terms as impairment) into account, and we recommend that new legislation be developed, building on this new evidence.

Finally, we examine the changing nature of human communication communities over time, pointing out that effective governance mechanisms vary according to the community's environment. We then suggest that techniques emerging elsewhere in the digital world, for example, assessment systems, trustworthiness indicators, and requirements for transparency, will be effective in today's communication environment.

## The Challenge

There are many efforts designed to protect teenagers using social media. Yet, imprudent, unethical, and bad behavior online—and the resulting harm among teenagers—continues to grow. Cyberexposure and cyberbullying are not eliminated or controlled by current social practices and laws and regulations.

Before addressing this challenge, we reach into enduring literature to remind ourselves that human beings—especially adolescents—have always been curious and engaged in sometimes-risky experiments and that there have always been bullies among us.

## What's New? Curiosity, Risk Taking, and Bullying Have Always Been with Us

Curiosity, risk taking, and bullying are not new. The often quoted proverb, *curiosity killed the cat*—documented as early as 1598 (Johnson, 1598/2009)— reminds us that natural curiosity is often seen as leading people into trouble.

Ponton (1998) asserted that risk taking is a normal and essential part of the teenage years but suggested that, unless positive risk taking is modeled, young people, especially those under stress, will often choose negative risks.

And bullies are also a staple of enduring fiction, as exemplified by "Bones and his gang of rough riders" in *The Legend of Sleepy Hollow* (Irving, 1820) and Horatio Alger's bully Micky Maguire and his gang in *Ragged Dick* (1868).

If these behaviors are not new, what has changed?

- First, there are important changes in the communication environment, which amplify adolescent visibility and vulnerability in the online global village.
- Second, there is new evidence regarding the timing of human brain development as it relates to the concept of diminished capacity, which is relevant to the laws regarding social media.
- Third, there is a new understanding of the ways in which effective forms of governance—ethical norms, laws, and other means of managing behavior—co-evolve with changes in society.

We discuss each of these areas in turn, beginning with the amplification effects of social media.

## The Amplifying Effects of Social Media

All communication processes and technologies have distinct characteristics that enable some activities and prevent others. These range from being able to see a person's body language in a face-to-face conversation to being able, in some cases, to track someone's location through a mobile phone's global positioning system and *not* being able, in other cases, to verify the identity of the person making some online postings.

Social media technologies create *amplifying effects*. Western society first experienced a major amplifying effect as the result of new communication technologies with the spread of the printing press in early Modern Europe (Eisenstein, 1983). Books, pamphlets, and fliers spread ideas much more quickly and extensively than when humans relied on word of mouth and laborious, handwritten documents. The next major amplification came with the introduction of electronic communication and the development of broadcast mechanisms in the first half of the 20th century. Radio, television, and films spread information to vastly larger audiences, reducing the time it took information to travel from years to weeks.

In the last 20 years, four new developments united to create another substantive increase in amplification. First, the cost and size of the equipment (e. g., mobile phones and tablets) needed to create and send messages dropped significantly and purchase of these items spread quickly. Second, multidirection networks were developed and widely deployed, including both wired and wireless access. Third, "front doors" to communication networks in the form of easy-to-use interfaces, utilities, and services such as social networking sites were introduced and widely adopted (for example, web browsers, e-mail and texting services, and social networking sites). And fourth, messages, images, and information began to be stored and transmitted in digital form on communication networks.

As with earlier communication revolutions, these characteristics have combined in such a way that they greatly increase the speed and distribution of messages. In addition, however, with this change, everyone—including adolescents—gained the ability to communicate with huge audiences and global networks: an unprecedented power for individuals.

The major amplifying characteristics of social media that affect teenagers are as follows:

1. *We see increased and easy access to communication technologies with wide distribution capabilities.* An adolescent with Internet access can easily post a picture or video taken with a mobile phone to YouTube or Facebook. Of the 93% of American teens who use the Internet (Salmund & Purcell, 2011, p. 4), 73% use social networking sites (Lenhart, Purcell, Smith, & Zickhur, 2010, p. 2). More than 60% (Cox Communications, 2009, p. 19) have posted pictures of themselves online.

2. *Digital media are searchable, easily replicable, readily scalable, and persistent.* Contrary to most prior human experience, the ease with which digital information online can be located, copied, and distributed makes it almost impossible to call back mistakes or errors in judgment. Items posted in digital media persist. Once information is posted or sent, message originators lose control. This is especially relevant for teens whose cognitive abilities to consider the potential consequences of their actions have not yet fully developed.

3. *Invisible audiences exist.* When adolescents post information online, they rarely have a clear conception of the size and scope of the potential audience for their information (boyd, 2008). Because the audience is invisible, it is difficult—if not impossible—to clearly

imagine, and, therefore, to take into account everyone who might gain access. Teens have neither the inclination, nor the ability, to monitor their audiences. Thus communication often takes place without regard to the potential consequences of wide distribution.

Not seeing one's audience face-to-face removes the ethical inhibitions that check potentially exposing or aggressive communication that human society has developed over millennia. Psychologist John Suler (n.d.) wrote the following:

Inhibiting guilt, shame, or anxiety may be features of the in-person self but not that online self (The Online Dishibition Effect).

Among the contributing psychological factors that Suler found are the following: dissociative anonymity (a feeling that others on the Internet won't know us), feelings of invisibility (a sense that others can't see us), asynchronicity (a feeling of disconnection in time), and solipsistic introjections (a sense that digital communication is not "real" in a practical sense).

4.  *There is global contact and anonymity*. Online, the world becomes McLuhan's global village. Using information developed by, communicating with, and buying from strangers becomes commonplace. While there are many benefits of extended contact, there are also dangers.

    Without physical presence, it is more difficult to identify when others are trustworthy. In many cases, people online or communicating via a mobile device can arrange to be anonymous, adopt a nonrepresentative presence, or impersonate someone else (Donath, 1999). Thirty-two percent of online teens have been contacted by a stranger. Fortunately, most are not upset by the contact (Lenhart, 2008, p. 12). But some stranger contact is harmful and, when this is the case, most individuals and families do not have the skills or resources to conduct a thorough investigation.

5.  *There is continuous immersion in mediated communication*. Ipsos Reid (2011) used the term *continuous users* to describe the rising percentage of teens (now more than a quarter of all teens) who describe themselves as being immersed in social media. Sixty-five percent of teens who can access the Internet via their mobile phones are continuous Facebook users. In addition, 68% of online teens and young adults use the Internet to get news and 48% have made online purchases. Thus mediated communication is coming to permeate all aspects of their lives.

The difficulty of escaping mediated harassment, bullying, and even exposure to untrustworthy and undesirable information (spam, ads, pornography, and inaccurate information) has greatly increased. Unlike face-to-face bullying, from which a young person can temporarily escape by going to places where there are trusted authorities, it is difficult to escape social media, especially at a time when peer interaction is essential to healthy development.

6. *The blurring of public and private affairs and communication with limited context.* With continuous communication comes the blurring of formerly strong boundaries; the differences between school or work and home, between public and private, and between formal and informal are all fading (Allison, 2005; boyd, 2008; Meyrowitz, 1985). This leaves young people to make distinctions and establish boundaries for themselves. Most electronically mediated information provides little context, which exacerbates this challenge.

7. *We see emerging norms and ethics.* Some scholars posit that communication behavior online and via mobile devices takes place in a normative vacuum (Finkelhor, 2010). It is true, of course, that norms and ethics lag behind innovation, as people must have experience with new technologies and processes before a norm or more formal governance, such as a law, can emerge. Nevertheless, we see emerging protocols and governance, for example, concerning the use of cell phones in public spaces or the informality and brevity of texts and tweets.

Rather than an ethical and normative void, we find evidence of a clash of norms and ethics from the Manufacturing Age with those emerging in the Information Age (Allison, 2005; boyd, 2008; Finkelhor, 2010). As might be expected, the emerging Information Age norms often scandalize more formal, traditional adults who are often authorities and legislators. As more adolescents use social media, some social power migrates from traditional authorities to those who use media more effectively.

In the United States, our primary form of governance rests with laws and the courts and police who enforce them. In the next section, we discuss the new evidence on brain development as it is related to the law.

# The Law and the Rate of Mental Development of Adolescents Using Social Media

In the Middle Ages, the Catholic Church considered a person to be an adult at the age of 7. The justification was that a 7-year-old was assumed to know the difference between right and wrong (Postman, 1994).

Times change, but the question of the age at which it is appropriate to treat a person as an adult is still not settled in the United States. As recently as 2001 in Florida, Nathaniel Brazill, 13 at the time of the crime and 14 at his trial, was convicted of second-degree murder and sentenced to 28 years in prison (*Brazill vs. Florida*, p. 1). In some other states, he would have been tried as a child or juvenile. The age at which one can be prosecuted as an adult has a long history of debate and change. The law for lesser offenses also fails to show consensus around juveniles.

It is no surprise, then, that Internet-mediated crimes would be subject to similar confusion. In January 2009, after a year of examining available research, the Internet Safety Technical Task Force (ISTTF) issued a report to the Multi-State Working Group on Social Networking of State Attorneys General of the United States. It concluded that cyberbullying and harassment are the most important risks that youth face. The law in these areas is in flux.

Forty-four states, as of January 11, 2011, had laws on the books specifically addressing bullying. Eleven were considering such laws. Six states have cyberbullying in their laws. Thirty-one states include "electronic harassment" in their statutes (Hinduja & Patchin, 2009, p. 3). Four states have changed their laws to reduce penalties for teenagers who have been charged with sexting as a criminal matter and 14 more are considering such changes to differentiate between teenagers and adult pornographers and sexual predators (NCLSL, 2011).

These problems stem from the same technology: access to instant and widespread communication. In the effort to control harm, two different approaches are appearing.

In the case of cyberbullying, existing laws are being used or adapted and new laws created to ensure that mechanisms exist to punish offenders and shield victims. Often this means creating laws specifically to allow effective prosecution of persons bullying others.

Sexting, like cyberbullying, is a peer-to-peer activity that does not involve adult predators or pornographers. Unfortunately, the two are often conflated (Gifford, 2008). Some commentators feel—and we agree—that the changes in law are coming neither fast enough nor do they take into account the differences between adult predators and young people experimenting with the combination of sexual development and new technology. For example, Szymialis wrote the following in the *Indiana Law Review* (2010):

> The law has failed to adapt quickly enough to teens sending these images. In response to sexting, prosecutors have utilized laws originally intended for child predators, such as child pornography statutes. Many of these statutes define the prohibited acts using broad language.
>
> For example, Pennsylvania's child pornography statute, the statute under which teens could be prosecuted for sending sext messages, prohibits depictions of minors "engaged in a prohibited sexual act." Nudity is included in the definition of a "prohibited sexual act" if the depiction is sexually stimulating. For parents and teens facing an aggressive prosecutor, this takes the phrase "in the eye of the beholder" to a whole new level. A conviction under a child pornography statute, "even in juvenile court," may require classification and registration as a sex offender for the juvenile. This registration includes "community notification requirements" (p. 302, paragraphing added).

A further complication to this restructuring of the law to accommodate the digital environment is the result of recent studies on the development of the brain. The use of high-resolution structural magnetic resonance imaging (MRI) has allowed researchers to observe the development of the brain in vivo. A 1999 study, in vivo evidence for postadolescent brain maturation in frontal and striatal regions (Sowell, Thompson, Holmes, Jernigan, & Toga, 1999) showed that—contrary to previous belief that the brain had largely finished development around the time of puberty—the brain actually continues to develop into the early 20s. Later studies conducted for the National Institute of Mental Health (NIMH, n.d., p. 1) confirmed these findings.

The NIMH (n.d.) and subsequent studies (Juvenile Justice Center [JJC], 2004) have demonstrated that the prefrontal lobe—specifically that part of the brain that deals with consequences—is the final part of the brain to develop. Subsequent refinements have raised the full-development age to between 20 and 25 (Luna, 2005). The director of the University of Pennsylvania Medical Center wrote that the evidence is strong that the brain does not mature until the early 20s in areas that govern "impulsivity, judgment, planning for the future, foresight of consequences, and other *characteristics that make people morally culpable...*" (Gur in JJC, 2004, p. 3, emphasis added).

Empirical evidence of this lack of capacity has been available for quite some time. Car rental firms, for instance, often refuse to rent cars to persons under 25. They do this because accident rates are much higher for renters under 25.

Nevertheless, it seems difficult for our society to apply this knowledge to our legal system. Society continues to insist that young people rationalize their poor behavior, rather than understanding that, in many cases, they are literally not capable of understanding the possible consequences of their actions. For example, in the 1999 *Handbook of Moral Behavior and Development,* referring to juveniles involved with the criminal justice system, Gibbs described young people as having

> a sociomoral development delay, that is, the persistence beyond childhood of egocentric bias that appeared to be supported by significant cognitive distortions of situations presented. (Willard, 1997, p. 6)

He went on to posit that young people develop rationalizations that externalize blame. These rationalizations

> appeared to protect the individual from considering the factors that might restrict inappropriate behavior, such as empathy for the victim or dissonance with self-concept. (Willard, 1997, p. 6)

Once we factor in the new information on brain development derived from MRIs, the issue becomes clear. Youthful offender explanations are not always rationalizations but sometimes the best explanations the offenders can come up with absent a fully developed frontal lobe.

This new information raises significant legal and ethical issues. *Diminished capacity* is already a generally accepted concept in American law:

> The diminished capacity plea is based in the belief that certain people, because of mental impairment or disease, are simply incapable of reaching the mental state required to commit a particular crime. In the example of murder and manslaughter, a diminished capacity defense contends that a certain defendant is incapable of intending to cause a death, and therefore must have at most caused such a death recklessly. Thus, a successful plea of diminished capacity in a murder trial would likely result in the charge being reduced to manslaughter. (Legal Information Institute, 2010)

The phrase *arrested or retarded development,* while not intended to apply to healthy, normal people, would appear to apply to those under 25, according to the MRI-based findings on the brain. How can we ethically hold youth who may not be *capable* of foreseeing consequences or restraining impulses to fully account in the same way that we would hold mature persons accountable?

It is important to note that diminished capacity is not, nor in our opinion should be, considered a defense that would lead to a dismissal of a charge. Instead, it is intended as mitigation to soften the sentence, often leading to treatment rather than confinement. In light of the new information, efforts to craft and modify laws to deal with cyberbullying, sexting, and other social media offenses should take into account that seeming adults younger than 25 are operating with brains that still lack the ability to fully understand consequences

As society changes, effective methods of governance also change. In the next section, we turn our attention to a review of historically effective and emerging governance mechanisms.

## The Changing Nature of Communication Communities

Effective governance changes with the type of communication community to be governed. By *communication community,* we mean the group with which an individual can readily communicate. Along with the other elements of society, communication communities change over time in a series of large steps or punctuations of evolution.

These evolutionary steps are often colloquially known by the characteristic human occupation of the time: the Agricultural Age, the Manufacturing or Industrial Age, and the (working with) Information Age. Because we want to focus not on the work performed but on the changing nature of the social systems, in this chapter, we use the terms developed by German sociologist Ferdinand Tönnies (1996) and extended by Mary Ann Allison (2005). Rather than Agriculture, we will use Tönnies's label of *gemeinschaft* to describe the social patterns of that time. Tönnies used *gesellschaft* to label society in the Age of Manufacturing and, to honor and extend the Tönnies convention, Allison has named the society emerging in the Information Age *gecyberschaft.*

Because societies and communities are complex adaptive systems, there is no single, linear cause-and-effect relationship. Factors triggering change affect each other as well. Nevertheless, it is useful to distinguish major elements. When considering adolescents using social media, there are five major characteristics that affect the communication communities in which they act:

1. Changes in the tools and processes humans use to communicate
2. Changes in the size of communities that are readily accessed
3. Changes in community communication norms
4. The pace of social and technological change
5. Changes in governance mechanisms that will be effective

## 1. Communication Tools and Processes

As discussed earlier in the section on amplification (and not repeated here), it is only within the last decade that adolescents have gained the capacity to broadcast messages on a global scale. In addition, recent Internet and mobile phone technologies often make information "instantly" available.

## 2. Size of Ready-Access Communities

Communication community size also changes over time. In *gemeinschaft* (the Agricultural Age), most people lived in villages with a population of 150 or less (Allison, 2005; Tönnies, 1996). People rarely traveled more than seven miles in their lifetime (Burke, 1999). In this time, then, adolescents had a communication community of 150 or less. Dunbar's research (2003; Dunbar & Hill, 2003) into the correlation between mammal brain size and the size of mammal groups leads us to the conclusion that adolescents, their families, and local authorities all knew everyone in the village. Visitors were rare and interaction with strangers was actively discouraged.

In the mid-19th century, triggered in part by the results of the spread of printed information, humans started building and migrating to cities in large numbers. This was a slow process. It wasn't until 2008 that half the world's population lived in cities (Martine, 2007). During the early stages of this shift to cities, Tönnies (1996), Durkheim (1960), and Weber (1947) noticed and documented a complete reshaping of society by life in urban areas. Drawing the clearest distinction, Tönnies compared life in villages (which he termed *community* or *gemeinschaft* in German) to life in cities (which he termed *society* or *gesellschaft* in German).

*Gesellschaft* city dwellers function in a new and larger communication community. While we can create social media and databases with thousands of contacts, human brains (Dunbar, 2003; Dunbar & Hill, 2003) do not permit us to know thousands of people well enough to determine whether they are, for example, trustworthy.

City population varies but, by definition, cities are larger than villages. During *gesellschaft*, adolescents typically did not have access to broadcast communication technology and so did not communicate with everyone in the city. Nevertheless, passing people on the sidewalk, meeting others in stores, and jostling next to each other in sports stadiums, adolescents had ready communication access to many more than 150 people.

Urban residents cannot help but encounter and communicate with strangers. Early cities were—and some parts of modern cities, especially rapidly growing cities are—dangerous. But over time, humans learned how to govern cities. While urban residents don't always dwell together peacefully, more than half of the world—including adolescents—have learned how to live and communicate in cities.

In the last decade, with the advent of widespread access to the World Wide Web, fronted by easy-to-use interfaces, utilities, and services in *gecyberschaft* (the Information Age), the *potential* communication community for adolescents has risen to more than two billion people (International Telecommunication Union [ITU], 2011, p. 4).

## 3. Community Communication Norms

Because the environment and the size of the communication community changes, the associated social norms change as well.

In *gemeinschaft*, communication norms were characterized by the intimacy that comes from face-to-face communication with people known to the communicator. Norms grew out of daily interaction; respect of authority was traditional and enforced physically. Trust was based on experience.

In *gesellschaft*, communication norms became more formal. People often communicated with others they did not know. Some forms of proper interpersonal communication were taught in the increasingly required schooling. Ownership of communication and information (copyrights and patents) became important. As radio, TV, and film broadcasting developed, laws regarding these powerful forms of communication were instituted. Concepts of fairness and objectivity were emphasized. Norms for receiving broadcasts—for example, "don't talk in movie theaters"—developed naturally. In this larger communication community, laws regarding interpersonal communication also became necessary.

Now, in *gecyberschaft*, communication norms are still emerging to help people handle a communication environment that is immersive, polychromic, and continuously mediated. In addition to the *gesellschaft* goal of objectivity, communities often call for multiple viewpoints. As demonstrated in the Creative Commons movement (Creative Commons, n.d.) and in scholarly papers where the single author paper is disappearing (Greene, 2011), authorship and ownership are moving toward a recognition of the networked groups that create and assemble messages and information. Informality in the global village among known correspondents mirrors in some ways the earlier informality in physical villages.

## 4. The Pace of Social and Technological Change

The pace of social and technological change has been accelerating throughout human history (Kurzweil, 2005; Laszlo, 1996; Smart, 2005). Matching the accelerating pace of change, the level of social complexity has also increased exponentially (Bar-Yam, 1997; Kaufman, 1995; Luhmann, 1982).

For individuals, this is reflected in the increasing number and expected speed of social interactions and in the escalating complexity of the information to be navigated. For society, this creates a parallel challenge: there is a need for governance processes that are capable of being in step with the complexities and speed of social and technological change.

In *gemeinschaft,* it took centuries for new technologies and processes to move through society. Most individuals lived where their parents and grandparents had lived and did what their parents and grandparents did (Tönnies, 1996). Communication had low intensity, and communication with "the outside world" was minimal and often nonexistent (Durkheim, 1960). Because there was little change, the wisdom of elders, reflecting past experience, sufficed to maintain order (Weber, 1947, 1968).

In *gesellschaft,* the speed of social and technological change increased. Communication increased in speed and intensity, and even in small villages there was frequent communication with the larger world (Durkheim, 1960). People were more mobile and began in significant numbers to live in cities. New technologies, the mixing of diverse peoples, and the increase in frequency and types of communication triggered an environment in which change took place over decades rather than centuries. The wisdom of elders was not always sufficient to address the new complexities or the faster pace of change (Weber, 1947, 1968).

The trend of a faster and more complex society and technology continues to accelerate, with a punctuation point taking shape over the last decade and resulting in the new social age, *gecyberschaft* (Allison, 2005), more colloquially called the Information Age. Not only is technology proliferating (Kurzweil, 2005), but the time it takes before a significant percentage of the population has adopted new technology is shrinking. It took 38 years after the radio was introduced for 50 million people to own one (Xplane, 2009). But it took TV sets only 13 years to reach 50 million people. And it was only 4 years after they became widely available that 50 million people had purchased laptops.

It is important also to take into account the pace of information development and the increasing immersion in communication. Computer scientists at

IBM (Coles, Cox, Mackey, & Richardson, 2006) estimate that the information accessible to humans and computers now doubles every 11 hours.

Adolescents are at the forefront of this acceleration. The typical American adolescent sends and receives more than 50 texts a day, with 31% of teens sending and receiving more than 100 messages a day (Lenhart, Ling, Campbell, & Purcell, 2010). Seventy-five percent of American adolescents have a Facebook page and 27% of them report that they access their Facebook page continuously throughout the day (Ipsos Reid, 2011).

This rapid pace creates uncertainty and stress. As early as 1989, Wurman (1989/2001) coined the phrase *information anxiety*. Social norms have difficulty keeping pace (Collier, 2010; Scott, Ferestad, & Ellestad, 2005).

Just as increasing the number of social connections an individual has (Dunbar, 2003; Dunbar & Hill, 2003), increasing the interconnections at the institutional level—global supply chains, linked financial markets, borderless media, travel and migration, among many others—adds to the challenge faced by those charged with keeping order (Deibert, 1997). The pace and complexity strains government bureaucracies—as they were structured for a slower, less connected, and complex time (Allison, 2005; Fukuyama, 2002; Kelly & Allison, 1999).

## 5. Effective Governance Mechanisms

In *gemeinschaft*, villages were governed by custom, monitored by elders, and enforced through social pressure. In a village of 150, everyone knew everyone else. It was difficult, if not impossible, to hide activities. Repeat offenders were shunned or excluded from the communication community. The ultimate sanction was banishment, with death the likely result.

As the Amish in the United States who live in *gemeinschaft* societies are finding, social pressure is not effective for communities of more than 150 (Nolt, 2004). Shunning doesn't work if everyone doesn't know everyone else, because one doesn't know who to shun. When banishment simply means the offender moves to the nearest town, it has little enforcing power. Larger Amish communities must turn to *gesellschaft* bureaucracy—frequently the police and courts—to maintain order (Robinson, Scaglion, & Olivero, 1994).

The widespread use of the printing press in early modern Europe (Eisenstein, 1983) was a major factor in accelerating technological and social change. Community size increased and even smaller groups, who often thought of themselves as traditional villages, had ever-increasing contact and commerce with "the outside world" (Vidich & Bensman, 1968). Isolation was no longer possible and *all* of society, at varying speeds, moved into *gesellschaft*. As Timmins (1998) demonstrated, the physical signs of village life—smaller populations, white picket fences, and grassy town squares—do not hold out the changes that came with telegraph-supported newspapers, radios, TV, films, and interstate commerce via trucks and trains. With larger populations in which to maintain order, *gesellschaft* societies developed bureaucracies: systems of law, police, and courts to govern people who were strangers to each other. Because society was so interconnected, regional and national authorities were also needed.

Now we are moving into *gecyberschaft*, where almost all human populations on the planet Earth are inextricably connected by communication, travel, and commerce. National laws, courts, and police do not have global jurisdiction. Just as social pressure is not sufficient in *gesellschaft*, bureaucratic laws, courts, and police are not sufficient in the Information Age.

Effective means of governance are still emerging. Among the many examples of new forms are the global marketplaces that connect individual buyers and sellers or employers and workers, such as eBay, Elance, and Mechanical Turk. Although respectful of local laws, these global organizations have developed assessments and dispute resolution mechanisms that enable a new form of practical trust among strangers around the world. The World Economic Forum (n.d.), which brings together heads of state and business leaders to address global concerns, and the World Trade Organization, with 153 nation-state members, which "deals with the rules of trade between nations at a global or near-global level" (WTO, n.d.), are examples of self-organizing systems working to address the governance of a connected world. Transparency— revealing information showing that there is nothing to hide—is another frequently emerging governance process (Lathrop & Ruma, 2010; Transparency International, n.d.).

One of the most interesting characteristics of social evolution is that the older social forms do not disappear but are, instead, incorporated—slightly modified—in the new age (Allison, 2005). For example, *gecyberschaft* villages continue in the form of our friends and family, where humans still respond to social pressure. Nations, with their laws, courts, and police, continue to function as well; but we are all—including adolescents—faced with many more cross-border and global connection challenges.

When faced with protecting and governing young people, it is essential that we consciously use all of the governance mechanisms already developed as well as those emerging in *gecyberschaft*.

# Addressing Amplified Adolescent Visibility in the Global Village

It makes common sense and is ethically important to understand a situation as thoroughly as possible before addressing it. We have explored three factors that are key to understanding and addressing adolescent participation in a mediated and connected world. The dangers are well known, perhaps even overhyped. Not so widely understood are the various forms and consequences of communication amplification, the slow development of the human brain, and the range of governing tools available and appropriate for this time in human history.

Believing that effective solutions will naturally arise from a more comprehensive understanding of the situation, our primary goal here was to identify and illuminate these points, not to suggest particular tactics. We conclude with a few short implications and invite readers to suggest others.

First, we suggest that the social conversation around young people and social media include a richer and more balanced view of the characteristics and consequences. If we aren't conscious of the distancing, boundary-blurring, and loss of control aspects of social media, for example, it is unlikely our responses will be effective.

Second, it is not effective or ethical to hold adolescents responsible when they are less capable of forecasting and impulse control than full adults. In a just society, laws and expectations reflect the actual situation, not what we might hope for. In addition, adolescent diminished intellectual capacity makes exhortation, education, and legislation useful but unlikely to be as effective as might be hoped in modifying cyberactivity. This brings us to our third point.

In addition to family and school guidance and updated laws, when seeking to protect adolescents, we hope that parents and authorities will consider what use may be made of emerging global governance mechanisms such as assessment systems, trustworthiness indicators, and requirements for transparency in, for example, matters of identity. The social media environment can, itself, be used to offer simulations and support structures to bolster teenage understanding of potential consequences.

Finally, there are good reasons for optimism. When people first started moving to cities in large numbers, they were often frightened of being in the midst of strangers. Crime was rampant. But over time the dominant mode—urban dwellers—developed *gesellschaft* society with laws, courts, and police, as well as norms and ethics, to maintain order and safety while enabling people to partake of the richness of cities. The British comedian and philosopher Stephen Fry summed up the evolution of learning, ethics, and governing mechanisms that the residents of *gecyberschaft* society are now developing to provide similar order and safety for everyone, including adolescents, while using the richness offered by social media.

But the Internet is a city and, like any great city, it has monumental libraries and theatres and museums and places in which you can learn and pick up information and there are facilities for you that are astounding—specialised museums, not just general ones.

But there are also slums and there are red light districts and there are really sleazy areas where you wouldn't want your children wandering alone....

What you don't need is a huge authority or a series of identity cards and police escorts to take you round the city because you can't be trusted to do it yourself or for your children to do it.

And I think people must understand that about the Internet—it is a new city, it's a virtual city and there will be parts of it of course that they dislike, but you don't pull down London because it's got a red light district (Fry, 2009).

## References

Alger, H. (1868). *Ragged Dick.* Boston, MA: A.K. Loring. Reproduced by the Electronic Text Center, University of Virginia Library. Retrieved from http://etext.virginia.edu/toc/modeng/public/AlgRagg.html

Allison, M. (2005). Gecyberschaft: A theoretical model for the analysis of emerging electronic communities. *Dissertation Abstracts International, 65*(11A), 4366. (UMI No. 3155725)

Bar-Yam, Y. (1997). *Dynamics of complex systems.* Reading, MA: Addison-Wesley.

boyd, d. (2008). *Taken out of context* (Unpublished doctoral dissertation). University of California, Berkeley. Retrieved from http://www.danah.org/

*Brazill v. Florida.* 845 So. 2d 282 - Fla: Dist. Court of Appeals, 4th Dist. 2003.

Burke, J. (1999). *The day the universe changed: Gutenberg's press.* New York: BBC-TV. Distributed by Ambrose Video Publishing.

Coles, P., Cox, T., Mackey, C., & Richardson, C. (2006). *The toxic terabyte: How data-dumping threatens business efficiency.* IBM Global Technology Services. Retrieved from http://www.conectividad.org/archivo/estudios/IBM_toxic_terabyte.pdf

Collier, A. (2010). *Social norming: *So* key to online safety.* NetFamilyNews.org. Retrieved from http://www.netfamilynews.org/?p=28726

Corwin Press. (2008). *High-tech bullies: What educators and parents can do.* Retrieved from http://www.corwin.com/repository/binaries/PressReleases/Hinduja-Patchin8-08.pdf

Cox Communications. (2009). *Teen online & wireless safety survey: Cyberbullying, sexting and parental controls.* Retrieved from http://www.google.com/url?sa=t&rct=j&q=&esrc=s&

source=web&cd=1&cad=rja&ved=0CCIQFjAA&url=http%3A%2F%2Fww2.cox.com%2
Fwcm%2Fen%2Faboutus%2Fdatasheet%2Ftakecharge%2F2009-teen-survey.pdf%3Fcamp
code%3Dtakecharge-research-link_2009-teen-survey_0511&ei=B5FTUL_CD6rg0QGOx
YCwCg&usg=AFQjCNH4yrCWDa8-YCeSgRin95fzcg3osg

Creative Commons. (n.d.) *Share, remix, reuse—legally.* Retrieved from https://creativecom
mons.org/

Deibert, R. (1997). *Parchment, printing, and hypermedia: Communication in world order transformation.* New York: Columbia University Press.

Donath, J. (1999). Identity and deception in the virtual community. In P. Kollock & M. Smith
(Eds.), *Communities in cyberspace* (pp. 27-58). London: Routledge.

Dunbar, R. (2003). The social brain: Mind, language, and society in evolutionary perspective.
*Annual Review of Anthropology, 32,* 162–181.

Dunbar, R., & Hill, R. (2003). Social network size in humans. *Human Nature, 14*(1), 53–72.

Durkheim, É. (1960). *The division of labor in society* (G. Simpson, Trans.). New York: Free Press.

Eisenstein, E. (1983). *The printing revolution in early modern Europe.* New York: Cambridge
University Pres.

Finkelhor, D. (2010, October 22). *The Internet, youth deviance and the problem of juvenoia.* Video
presentation at the University of New Hampshire, Durham.

Fry, S. (2009). Stephen Fry: The internet and me. *BBC News.* Retrieved from
http://news.bbc.co.uk/2/hi/7926509.stm

Fukuyama, F. (2002). *The great disruption: Human nature and the reconstitution of social order.* New
York: Touchstone.

Gifford, N. (2008). *Trends and prevention.* The Family Online Safety Institute. Retrieved from
http://www.cbhcare.com/files/u2/Trends_and_Prevention-Nancy_Gifford.pdf

Greene, M. (2011). The demise of the lone author. *History of the Journal* Nature. Retrieved from
http://www.nature.com/nature/history/full/nature06243.html

Hinduja, S., & Patchin, J. (2009). *Cyberbullying fact sheet: What you need to know about online
aggression.* The Cyberbullying Research Center. Retrieved from www.cyberbullying.us/cyber-
bullying_fact_sheet.pdf

Hinduja, S., & Patchin, J. (2010). *Cyberbullying: Identification, prevention, and response.* The
Cyberbullying Research Center. Retrieved from http://www.cyberbullying.us/Cyberbullying
_Identification_Prevention_Response_Fact_Sheet.pdf

International Telecommunication Union (ITU). (2011). *The world in 2010: ICT facts and figures.* Geneva, Switzerland: Author. Retrieved from http://www.itu.int/ITU-D/ict/statis
tics/index.html

Internet Safety Technical Task Force (ISTTF). (2009). *Enhancing child safety and online technologies: Final report of the Internet Safety Technical Task Force to the Multi-State Working Group
on Social Networking of State Attorneys General of the United States.* Retrieved from
http://cyber.law.harvard.edu/research/isttf

Ipsos Reid. (2011). *One quarter (27%) of American teens use Facebook continuously throughout the
day.* Retrieved from http://www.ipsos-na.com/news-polls/pressrelease.aspx?id=5095

Irving, W. (1820). *Rip Van Winkle & the legend of Sleepy Hollow: Found among the papers of the
late Diedrich Knickerbocker.* The Harvard Classics Shelf of Fiction. Retrieved from http://www.
bartleby.com/310/2/2.html

Jonson, B. (1598/2009). *Every man in his humour.* Project Gutenberg. Retrieved from http://www. gutenberg.org/files/3694/3694-h/3694-h.htm

Juvenile Justice Center (JJC). (2004). Adolescence, brain development, and legal culpability. American Bar Association. Retrieved from http://www.google.com/url? sa=t&rct=j&q=& esrc=s&source=web&cd=1&ved=0CCIQFjAA&url=http%3A%2F%2Fwww.american-bar.org%2Fcontent%2Fdam%2Faba%2Fpublishing%2Fcriminal_justice_section_newsletter%2Fcrimjust_juvjus_Adolescence.pdf&ei=QPBUUIewKYHo0QHa9oCgCA&usg=AFQjCNHw5v3ODwD-_rQ5KIOi_6rnOedupA&cad=rja

Kaufman, S. (1995). *At home in the universe: The search for laws of self-organization and complexity.* New York: Oxford University Press.

Kelly, S., & Allison, M. (1999). *The complexity advantage: How the science of complexity can help your business achieve peak performance.* New York: McGraw-Hill.

Knox, R. (2010). *The teen brain: It's just not grown up yet.* National Public Radio. Retrieved from http://www.npr.org/templates/story/story.php?storyId=124119468

Kurzweil, R. (2005). *The singularity is near: When humans transcend biology.* New York: Viking Press.

Laszlo, E. (1996). *Evolution: The general theory.* Cresskill, NJ: Hampton Press.

Lathrop, D., & Ruma, L. (2010). *Open government: Collaboration, transparency, and participation in practice.* Cambridge, MA: O'Reilly Media.

Legal Information Institute. (2010). *Diminished capacity.* Cornell University Law School. Retrieved from http://topics.law.cornell.edu/wex/diminished_capacity

Lenhart, A. (2008). *Teens, online stranger contact and cyberbullying: What the research is telling us....* Pew Internet & American Life Project. Retrieved from http://www.slideshare.net/Pew Internet/teens-online-stranger-contact-cyberbullying?type=powerpoint

Lenhart, A. (2009). Teens and sexting: How and why minor teens are sending sexually suggestive nude or nearly nude images via text messaging. Pew Internet & American Life Project. Retrieved from http://www.pewinternet.org/Reports/2009/Teens-and-Sexting.aspx

Lenhart, A., Ling, R., Campbell, S., & Purcell, K. (2010). *Teens and mobile phones.* Pew Internet & American Life Project. Retrieved from http://www.pewinternet.org/Reports/2010/Teens-and-Mobile-Phones.aspx

Lenhart, A., Purcell, K., Smith, A., & Zickuhr, K. (2010). Social media & mobile Internet use among teens and young adults. Retrieved from http://www.pewinternet.org/Reports/2010/Social-Media-and-Young-Adults.aspx

Luhmann, N. (1982). *The differentiation of society* (S. Holmes & C. Larmore, Trans.). New York: Columbia University Press.

Luna, B. (2005). *Brain and cognitive processes underlying cognitive control of behavior in adolescence.* Invited speaker at the NIDA supported symposium of the AACAP/CACAP joint annual meeting, Toronto, Ontario, Canada.

Martine, G. (2007). *State of the world population: Unleashing the potential of urban growth.* New York: United Nations Population Fund. Retrieved from http://www.unfpa.org/swp/swpmain.htm

McLuhan, M., & Powers, B. (1992). *The global village: Transformations in world life and media in the 21st century.* New York: Oxford University Press.

Meyrowitz, J. (1985). *No sense of place: The impact of electronic media on social behaviour.* New York: Oxford University Press.

National Conference of State Legislators (NCSL). (2011). *2010 legislation related to "sexting."* Retrieved from http://www.ncsl.org/?tabid=19696

National Institute of Mental Health (NIMH). (n.d.) *Teenage brain: A work in progress (fact sheet)* (NIH Publication No. 01–4929). Retrieved from http://wwwapps.nimh.nih.gov/health/publications/teenage-brain-a-work-in-progress.shtml

Nolt, S. (2004). *A history of the Amish.* Intercourse, PA: Good Books.

Ponton, L. (1998). *The romance of risk: Why teenagers do the things they do.* New York: Basic Books.

Postman, N. (1992). *Technopoloy: The surrender of culture to technology.* New York: Alfred A. Knopf.

Postman, N. (1994). *The disappearance of childhood.* New York: Vintage/Random House.

Robinson, C., Scaglion, R., & Olivero, J. (1994). *The evolution of the police function in society.* Westport, CT: Greenwood Press.

Salmund, K., & Purcell, K. (2011). *Trends in teen communication and social media use: What's really going on here?* Washington, DC: Pew Internet & American Life Project. Retrieved from http://www.pewinternet.org/Presentations/2011/Feb/PIP-Girl-Scout-Webinar.aspx

Scott, F. O., Ferestad, J., & Ellestad, J. (2005, August). *Just say no: Exploring the norms of cell phone use in public.* Paper presented at the annual meeting of the American Sociological Association, Philadelphia, PA.

Smart, J. (2005, November). *How to be a strategic futurist: An evolutionary developmental perspective on accelerating change.* Paper presented at the Tamkang University International Conference: Graduate Institute of Futures Studies, Global Soul, Global Mind, Global Action. Retrieved from http://www.accelerating.org/slides.html

Sowell, E., Thompson, P., Holmes, C., Jernigan, T., & and Toga, A. (1999). In vivo evidence for post-adolescent brain maturation in frontal and striatal regions. *Nature, 2*(10), pp. 859–861. Retrieved from www.loni.ucla.edu/~esowell/nn1099_859.pdf

Suler, J. (n.d.) The online disinhibition effect. *The psychology of cyberspace.* Retrieved from http://users.rider.edu/~suler/psycyber/disinhibit.html

Szymialis, J. (2010). Sexting: A response to prosecuting those growing up with a growing trend. *Indiana University Law Review, 44,* (X), pp. 301–339.

Timmins, J. (1998). When the mail-plane flies over: Space, civic identity, and small-town society. Written for Professor Mary Ryan's seminar on *Civic spaces in North America,* autumn 1996. Retrieved from http://www.geocities.ws/CollegePark/Campus/6925/papers/towns_society.htm

Tönnies, F. (1996). *Community and society* (C. Loomis, Trans.). New Brunswick, NJ: Transaction Publishers.

Transparency International. (n.d.). *The global coalition against corruption.* Retrieved from http://www.transparency.org/

Vidich, A., & Bensman, J. (1968). *Small town in mass society: Class, power and religion in a rural community.* Princeton, NJ: Princeton University.

Weber, M. (1947). *The theory of social and economic organization* (A. Henderson & T. Parsons, Trans.). New York: Free Press.

Weber, M. (1968). *Max Weber on law in economy and society* (Max Rheinstein, Ed., & Edward Shils & Max Rheinstein, Trans.). New York: Simon & Schuster.

Willard, N. (1997). *Moral development in the information age.* Center for Advanced Technology in Education, University of Oregon College of Education. Retrieved from http://tigger.uic.edu/~lnucci/MoralEd/articles/willard.html

Wolak, J., Mitchell, K., & Finkelhor, D. (2007). Does online harassment constitute bullying? An exploration of online harassment by known peers and online-only contacts. *Journal of Adolescent Health, 41*, S–51-S58.

World Economic Forum, The. (n.d.). Retrieved from http://www.weforum.org/

Wurman, R. (1989/2001). *Information anxiety 2.* Indianapolis, IN: Que.

Xplane. (2009, September 14). *Did you know? 4.0.* Retrieved from www.youtube.com/watch?v=6ILQrUrEWe8&feature=related

# · 6 ·

# Protection or Prosecution

## Julian Assange and Wikileaks
## Making Waves with a Cybersplash

ADRIENNE HACKER-DANIELS

In a prescient comment in response to a student's query in the wake of WikiLeaks' cyberdisclosure on July 25, 2010, of approximately 92,000 documents (Leigh & Harding, 2011, p. 117), the preponderance of which were classified, and all circumscribing the war in Afghanistan, Supreme Court Justice Sonia Sotomayor said that this event was likely to precipitate a judicial ruling on the relationship between and the balancing of national security and freedom of speech (Batty, 2010). If the legal exigence seemed compelling then, how much more compelling the exigence would become when WikiLeaks would post 391,832 classified entries, titled "The Iraq War Logs," on October 22, 2010 (Leigh, 2010b), culminating in, arguably, the most brazen posting, of over 250,000 State Department diplomatic cables, on November 28, 2010 (Leigh & Harding, 2011, p. 135).

WikiLeaks is the brainchild of Julian Assange, launched in 2006, with http://wikileaks.org registered as a domain name on October 4, 2006, posting as its first publication in December 2006, a "'secret decision,' signed by Sheikh Hassan Dahir Aweys, a Somali rebel leader for the Islamic Courts Union…" (Leigh& Harding, 2011, p. 56).

All three branches of government responded to WikiLeaks' document dumping. Members of Congress were quick to weigh in after the Cablegate dump. Congressman Peter King (R-N.Y.) exhorted that "WikiLeaks should be designated as 'a foreign terrorist organisation [sic],'" and that "WikiLeaks presents a clear and present danger to the national security of the US" (Leigh & Harding, 2011, p. 202). Secretary of State Hillary Clinton excoriated the November 28, 2010, dumping of confidential diplomatic cables in remarks aired Nov. 29, 2010, on PBS NewsHour, saying, "So let's be clear. This disclosure is not just an attack on America's foreign policy interests. It's an attack on the international community—the alliances and partnerships, the conversations and negotiations that safeguard global security and advance economic prosperity" (U.S. 'Deeply Regrets' Embarrassment of WikiLeaks Documents, Nov. 29, 2010). Secretary Clinton added that the leaks "in addition to endangering particular individuals, disclosures like these tear at the fabric of the proper responsible government" (U.S. 'Deeply Regrets' Embarrassment of WikiLeaks Documents, Nov. 29, 2010). And although President Obama has been uniquely aggressive in "pursuing leak prosecutions," (Shane, 2010) in his first public statement on the heels of the Afghanistan War Logs leak, on July 25, 2010, his ostensibly contradictory response serves as a portent to the predominantly legal as well as ethical issues that circumscribe WikiLeaks' actions and the concomitant fallout:

> While I'm concerned about the disclosure of sensitive information from the battlefield that could potentially jeopardize individuals or operations, the fact is, these documents don't reveal any issues that haven't already informed our public debate in Afghanistan. Indeed, they point to the same challenges that led me to conduct an extensive review of our policy last fall. (de Nies & Miller, 2010)

As a newly fashioned publishing model, Web 2.0, with its eradication of the author-reader dichotomy, was, in essence, a democratization of communication (Choate, 2008, p. 7). As print media were—in simplified parlance—construed as fundamentally reportorial modalities of communication, Web 1.0, originally conceived as a repository for content, now has as its sphere of influence, the strokes of the keyboard, contributing to the creation of reality in foundationally epistemological and ontological ways (Choate, 2008, p. 8). The proliferation of Web 2.0 social media like Twitter, YouTube, and WikiLeaks as social media platforms, has engendered new legal questions and ethical dilemmas regarding their capabilities to inform, persuade and galvanize individuals within the larger public arena, to create not only a new *modus operandi*, but a concomitant *modus vivendi* as well. If the "first generation of digital democracy" (Loader & Mercea, 2012, p. 1) left a lot to be desired, "...a fresh wave of technological optimism has more recently accompanied the advent of social media platforms such as Twitter, Facebook, YouTube, wikis and the blogosphere" (Loader & Mercea, 2012, p. 2). The cachet of this iteration is explained by Loader and Mercea (2012):

> The distinctiveness of this second generation of internet democracy is the displacement of the public sphere with that of a networked citizen-centered perspective providing opportunities to connect the private sphere of autonomouspolitical identity to a multitude of chosen political spaces.

> It thus represents a significant departure from the earlier restricted and constrained formulations of rational deliberation with its concomitant requirement for dutiful citizens. In its place is a focus upon the role of the citizen-user as the driver of democratic innovation through the self-actualised networking of citizens engaged in lifestyle and identity politics.

And although they credit WikiLeaks as a social media platform designed for the disclosure of government secrets (p. 2), the site and its raison d'être circumscribe more complex actions and underlying motives. Loader and Mercea highlight WikiLeaks' "disruptive capacity" along with its "potential power of collaborative sharing" (p. 5):

> Different in style from earlier forms of civic participation, such disruption is effected by enabling citizens to critically monitor the actions of governments and corporate interests. It could potentially enable political lifestyle choices to be informed through shared recommendations from friends, networked discussions and tweets, and direct interaction with conventional and unconventional political organisations (p. 5).

# "You Say You Want a Revolution"

The question of social media's ability to facilitate political change is addressed in the volume of essays, *The Global Impact of Social Media* (Bryfonski, 2012). Whether social media are facilitative is one question, and whether they are *expeditiously* facilitative is another. In "Social Media Fail to Incite True Activism," Malcolm Gladwell (2012) examines the activism employed by the four African American college students who dared to violate the discriminatory segregation laws and requested service at the Woolworth's lunch counter in Greensboro, North Carolina, on February 1, 1960 (2012, p. 95. For Gladwell, social media platforms engage a low threat activism, "built around weak ties" (2012, pp. 99–100). And as counterintuitive as it may sound, as conducive as social networks are to "…increasing *participation*," they do so "by lessening the level of motivation that participation requires" (2012, p. 101).

> Facebook activism succeeds not by motivating people to make a real sacrifice but by motivating them to do things that people do when they are not motivated enough to make a real sacrifice. We are a long way from the lunch counters at Greensboro (Gladwell, 2012, p. 101).

It would smack of shortsightedness to commit the fallacy of the false dilemma, for the "speech" associated with the ethereal communication in the clouds of cyberspace can be deployed symbiotically with the channel richness of the proverbial boots on the ground. For Gladwell, the lack of a hierarchical structure within social media, so prominent within more low-tech activism, is its Achilles heel (2012, p. 104). Some argue that the claims made as to the influence of social media in effectuating change is—at its fallacious extreme—a *post hoc* argument, and more magnanimously—an unknowable (Morozov, 2012, p. 107; Sharma, 2012, pp. 116–118). In "The Tunisian Revolt Was the World's First Facebook Revolution," Anshel Pfeffer (2012) relates the narrative of Sufian Belhaj, who as Hamadi Kaloutcha online, was inspired to translate documents about Tunisia which just 2 months prior had been published as part of the Cablegate cache of documents leaked by WikiLeaks (p. 92; Leigh & Harding, 2011, pp. 252–256). Belhaj said, "Of course the citizens of Tunisia already knew these things, but this made it official.…It gave frustrated people information that appeared to be reliable in place of whispered rumors" (p. 92). Daniel Arap Moi, former president of Kenya, accused of corruption, provided "…WikiLeaks [with] its first journalistic coup" (Leigh & Harding, 2011, p. 57). As reported, "The result was indeed sensational. There was uproar, and Assange was later to claim that voting shifted 10% in the subsequent Kenyan elections" (Leigh & Harding, 2011, p. 58). In his essay, "Social Media Empower People," Simon Mainwaring (2012) says the following of social media:

> As more people use social media to tell the story of the future, the wants and needs of more people will be reflected. Like all technology, social media is neutral but is best put to work in the service of building a better world. (p. 90)

But one might ask at what price—especially in the case of WikiLeaks. Assange's self-adulation might need to be tempered, as contemplated by Bill Keller (2012), editor of the *New York Times* at the time when the newspaper published many of the WikiLeaks documents (Keller, 2011). In reflecting on the WikiLeaks influence, Keller concludes that WikiLeaks really was not transformative, and that "it was a hell of a story and a wild collaboration, but it did not herald, as the documentarians yearn to believe, some new digital age of transparency" (Feb. 19, 2012). In fact, Keller argues that the effect might have been just the opposite:

> The publication of so many confidences and indiscretions did not bring U.S. foreign policy to a halt. But it did, at least temporarily, complicate the lives of U.S. diplomats. American officials say that foreign counterparts are sometimes more squeamish about speaking candidly, and that it is harder to recruit and retain informants around the world (Feb. 19, 2012).

# Genesis of a Cloak-and-Dagger Cyberstory

Julian Assange—at the epicenter of this massive cyberquake, was born July 3, 1971, in Queensland, Australia. By 1991, Assange was arguably "Australia's most accomplished hacker" (Leigh & Harding, 2011, p. 42). Within the throes of a turbulent familial life, including custody battles, Assange was working as a computer programmer, but he was far from gainful employment, given the lack of remuneration for his services (Leigh & Harding, 2011, p. 45). Leigh and Harding (2011) note that "Assange's early commitment to free information, and free software, would slowly evolve into WikiLeaks" (p. 45). Assange registered the domain name of http://wikileaks.org in 1999 as a "leakers website," but he let it lie dormant (Leigh & Harding, 2011, p. 46).

In one of the most illuminating profiles of a consummately elusive Assange, Raffi Khatchadourian (2010) delineates the provenance of Assange's thinking, culminating in the creation of WikiLeaks, in what Khatchadourian calls a "manifesto of sorts," titled, "Conspiracy of Governance." According to Khatchadourian, "Assange wrote that illegitimate governance was by definition conspiratorial—the product of functionaries in 'collaborative secrecy, working to the detriment of a population.'" "[Assange] argued, that, when a regime's lines of internal communication are disrupted, the information flow among conspirators must dwindle, and that, as the flow approaches zero, the conspiracy dissolves. Leaks were an instrument of information warfare" (Khatchadourian, 2010). Assange also said "...that his mission is to expose injustice, not to provide an even-handed record of events," (Khatchadourian, 2010) continuously in dogged and relentless pursuit of "illegal or immoral behavior" within oppressive regimes and corporations. Integral to WikiLeaks' raison d'être, Assange said the following:

Since unjust systems, by their very nature, induce opponents, and in many places barely have the upper hand, mass leaking leaves them exquisitely vulnerable to those who seek to replace them with more open forms of governance. Only revealed injustice can be answered; for man to do anything intelligent he has to know what's actually going on (Leigh & Harding, 2011, pp. 46–47).

And in a reference to Mark Felt of Watergate notoriety, in 2007 Assange made his first foray into a bona fide public arena with a statement reported by the CBC News in Canada, one of just a handful of media outlets to report the following:

> Deep Throat may be moving to a new address—online. A new website that will use Wikipedia's open-editing format is hoping to become a place where whistleblowers can post documents without fear of being traced. WikiLeaks, according to the group's website, will be "an uncensorable version of Wikipedia for untraceable mass document leaking and analysis. Our primary interests are oppressive regimes in Asia, the former Soviet bloc, sub Saharan Africa and the Middle East, but we also expect to be of assistance to those in the west who wish to reveal unethical behavior in their own governments and corporations" (Leigh & Harding, 2011, pp. 47–48).

# From Transmutations to Transmogrification: The Wiki-Wikipedia-WikiLeaks Continuum

If Wikipedia was the cyberintermediary, what is WikiLeaks' nascency? Ward Cunningham, whose soubriquet is the "father of wikis," launched the first wiki in March 1995 (Choate, 2008, p. 2), and after 2 months, he solicited participation from colleagues (Choate, 2008, p. 2). As told by Choate, *wiki*, Hawaiian for quick, linguistically captured Cunningham's goal, "...to use the World Wide Web to develop a way for programmers to more readily share ideas about design patterns" (p. 2). As a content management system (Choate, 2008, p. 2), wikis eradicated the author-reader dichotomy (Choate, 2008, p. 1), which was so entrenched legally, historically and culturally. Choate (2008)—convinced of the quickness of wikis—was unconvinced of their efficacy, until having witnessed the success of Wikipedia (p. 3). Wikipedia's cyberprogenitor, Nupedia.com, was founded by Jimmy Wales in March 2000, "as a peer-reviewed encyclopedia with a seven-step editing process" (Choate, 2008, p. 3). However, the process was unwieldy with a deficient yield; Choate provides the following account:

> As a solution to this problem, Editor in Chief Larry Sanger proposed a "feeder" site to Nupedia based on wiki technology on January 10, 2001. The idea was that people could post articles on the wiki and after those articles had been properly vetted, they could be moved onto Nupedia. The use of a wiki would make it easier for users to contribute

and, it was hoped, speed up the process. There was never any expectation at the time that Wikipedia would replace Nupedia, although this is what quickly happened. (p. 3)

In its first year, Wikipedia amassed 18,000 articles, by 2003 over 160,000 articles, and by the most recent count, as documented by Wikipedia, it currently boasts 3,892,865 articles in English alone (Wikipedia, n.d.). Wikis are classified under the rubric of Web 2.0 or social media, as steeped as those phrases are in semantic vagueness. Notwithstanding this vagueness, the list of uses renders the phrases comprehensible (Choate, 2008, p. 4). According to Choate (2008), wikis and other forms of social media share four attributes, insofar as they are all "participatory, decentralized, linked and emergent" (p. 4). As explained by Leigh and Harding (2011), Wikileaks has its communicative roots in wikis, having been conceived as "...a user-editable site..." (p. 52). But the irony of what Assange felt he needed to do should not go unnoticed:

> [when] Assange and his colleagues rapidly found that the content and need to remove dangerous or incriminating information made such a model impractical. Assange would come to revise his belief that online "citizen journalists" in their thousands would be prepared to scrutinize posted documents and discover whether they were genuine or not (Leigh & Harding, 2011, p. 52).

## ENTER BRADLEY MANNING:
## "Means, Opportunity and Motive"

In the fall of 2009, Bradley Manning, a specialist with the 2nd Brigade Combat team, 10th Mountain Division, was sent to Iraq, and given his aptitude with computers from an early age, he served as an intelligence analyst, privy to a voluminous amount of top secret information (Leigh & Harding, 2011, p. 20). With minimalist security and oversight, Manning was given two military laptops, providing significant means. Manning's *modus operandi*, for his role "[in] the largest leak of military and diplomatic secrets in US history" (Leigh & Harding, 2011, p. 22), was, by all accounts, incredible in its simplicity. Opportunity for Manning was not in short supply either. He allegedly used his rewritable Lady Gaga CDs, erased them, and copied the secrets onto them (Leigh & Harding, 2011, pp. 22, 31). His opportunity—given the surprisingly lax security—reached incipient fruition, when "his eye was caught by an exercise run by WikiLeaks on Thanksgiving 2009..." (Leigh & Harding, 2011, p. 31), in which WikiLeaks published over 500,000 pager messages intercepted

September 11, 2001, which had been obtained anonymously from a National Security Database (Leigh & Harding, 2011, p. 31).

But what were Manning's motives? Manning was disillusioned by the "official duplicity and corruption of his own country" (Leigh & Harding, 2011, p. 22), as evidenced by "videos that showed the aerial killing from a helicopter gunship of unarmed civilians in Iraq...[and] chronicles of civilian deaths and 'friendly fire' disasters in Afghanistan" (Leigh & Harding, 2011, p. 22). Manning also perceived the diplomatic cables from Cablegate as scandalous. Manning, like Assange, espoused the inviolability of access to free information in a democratic society (Leigh & Harding, 2011, p. 30). In a web chat initiated by Manning, with Adrian Lamo, a former hacker who was profiled in Wired.com, and who ultimately gave the tipoff to the FBI, resulting in Manning's arrest on May 26, 2010, Manning claims responsibility for the document and video leaks to WikiLeaks (Poulsen & Zetter, 2010). Manning was detained on May 26, 2010, by Army authorities (Poulsen & Zetter, 2010), with 22 subsequent charges, including 5 counts in violation of 18 U.S. Code Section 641, 8 counts in violation of 18 U.S. Code Section 793 (e), and 2 counts in violation of 18 U.S. Code Section 1030 (a) (1) (Department of Defense, Charge Sheet, July 5, 2010; Department of Defense Charge Sheet, Mar. 1, 2011). These charges encompass a range of actions resulting in the historic leaks of government data delineated below. On April 5, 2010, Wikileaks published a classified video titled, "12 JUL 07 CZ ENGAGEMENT ZONE 30 GC ANYOBE-avi (Poulsen & Zetter, 2010).The video, which WikiLeaks titled, "Collateral Murder," showed army helicopters firing down in a Baghdad suburb on a group of men, wounding children, and when it was all over, 12 people would be dead, including two employees of Reuters news agency, Namir Noor-Eldeen, a war photographer, and Saeed Chmagh, a driver and assistant (Leigh & Harding, 2011, p. 66).

1.  On July 25, 2010, WikiLeaks published the Afghan War Logs—its first foray into collaboration with traditional print media, having given "exclusive access" to two daily newspapers, the *New York Times* and the *Guardian*, and *Der Spiegel*, the weekly German magazine (Stelter & Cohen, Apr. 26, 2011). Over 92,000 documents comprise the Afghan War Logs, circumscribing "the failing war in Afghanistan" (Davies & Leigh, 2010).The evidence is damning, including the killing of at least 195 civilians and the wounding of at least 174 (Davies & Leigh, 2010).

2.  On October 22, 2010, WikiLeaks published the Iraq War Logs, the database of over 390,000 field reports quantifying the death and

destruction with numerical starkness. According to Leigh and Harding (2011), "the war logs detailed 109,032 deaths," including 66,081 civilian deaths, 15,196 Iraqi security forces deaths, and 23,984 deaths of persons classified as "enemy" (p. 130).

3. But arguably, and with considerable consensus, the most (ostensibly) damaging and most assuredly humiliating event was the coordinated leak of over 250,000 U.S. State Department cables. Although this unprecedented leak contained no top secret material (Leigh & Harding, 2011, p. 181), the cables did reach the level of SECRET NOFORN—meaning that non-U.S. citizens were never to lay eyes on this material (Leigh, 2010a). Joining the original mainstream media consortium (*New York Times, Guardian, Der Spiegel*), which released the War Logs, was France's *Le Monde* and Spain's *El Pais*. Notwithstanding some potentially catastrophic timing glitches in realizing a wholly synchronized disclosure for "maximizing political impact," (Leigh, 2010a; Leigh & Harding, 2011, pp. 194–199), Cablegate—as it would be known—had implications in terms of the parameters for first amendment rights, including most prominently, freedom of speech and freedom of the press, as they interface with media ethics and concomitant normative ethical perspectives.

Notwithstanding the redacting of names and editing of sensitive material prior to publication (Leigh & Harding, 2011, p. 190), the administration felt more acutely vulnerable with the simultaneous disclosure of these leaks in print and cyberspace than any of the others. Bill Keller, the editor of the *New York Times*, offered an apologia of sorts when he said the following:

> The cables tell the unvarnished story of how the government makes its biggest decisions, the decisions that cost the country most heavily in lives and money. They shed light on the motivations—and, in some cases, the duplicity—of allies on the receiving end of American courtship and foreign aid. They illuminate the diplomacy surrounding two current wars and several countries, like Pakistan and Yemen, where American military involvement is growing. As daunting as it is to publish such material over official objections, it would be presumptuous to conclude that Americans have no right to know what is being done in their name. (Leigh & Harding, 2011, p. 200)

The First Amendment issues circumscribing the WikiLeaks events are manifold—both manifestly and subtly.

# Espionage Act of 1917—
# Clear and Present Danger

Although the Justice Department honed in on a potential indictment of Julian Assange, using the Espionage Act of 1917, Attorney General Eric Holder acknowledged that solely using the Espionage Act is problematic (Savage, 2010). But the allegation of espionage is the starting point, originally legislated within the Alien and Sedition Acts of 1798, "enacted by the Federalist in order to strike back at the French…" (Tedford & Herbeck, 2009, p. 30). Sedition surfaced again upon the United States' entry into World War I with the passage of the Espionage Act of 1917. Section 793 of 18 U.S.C. delineates the crimes involving the "gathering, transmitting, or losing [of] defense information" (18 U.S.C. §793 (a-g))while Section 794 of 18 U.S.C. delineates crimes involving the "gathering or delivering [of] defense information to aid foreign governments" (18 U.S.C. §794 (a) (b) (c) (d). As Mary-Rose Papandrea (2008) explains in her superb article, "Lapdogs, Watchdogs, and Scapegoats: The Press and National Security Information," although 794 "prohibits disclosure to an 'agent…[of a] foreign government,.'…as with §793, it is quite possible to construe certain provisions of this statute to apply to the media…" (p. 268). And, although the word "communicates" is used in subsection (a), the word "publishes" appears only in subsection (b) (Edgar & Schmidt, 1973, p. 944; Papandrea, 2008, p. 268).

Ambiguity still abounds regarding the press's First Amendment protection, the sticking and tipping points being the intentionality of and the motives behind the publishing act, as well as the act's intended audience—a foreign enemy or the general public (Papandrea, 2008, p. 269). Edgar and Schmidt (1973) argue that Section 793 "raises substantial issues as to whatever much of the flow of defense information between executive branch employees and the press constitutes serious criminal defenses" (p. 967).The more hermeneutical issue seems to be "whether newspapers, their reporters, their informants, or anyone who investigates, accumulates, informs about, or retains defense information as a prelude to public speech is covered by the section" (p. 967). While subsections (a), (b) and (c) of Section 793 apply to pre-publication activities (Papandrea, 2008, p. 264), subsections (d) and (f) apply to the dissemination by those who lawfully possess the information (Papandrea, 2008, p, 264), and subsection (e) "prohibits the dissemination or retention…by those in 'unauthorized possession' of it (Papandrea, 2008, p. 264), rendering the media most vulnerable to the application of subsection (e) (Papandrea, 2008, p. 264).

Section 798—"Disclosure of Classified Information"—was enacted in 1950, contemporaneous with Section 793 (d) and (e), (Edgar & Schmidt, p. 1064) and applicable to anyone who "publishes" national defense information (Papandrea, 2008, p. 269). The most provocative tenet of Section 798 vis-à-vis the WikiLeaks activities is the ever-in-flux, contextually bound application of what constitutes "classified" information. In subsection (b) of Section 798, "classified" information is "information which, at the time of a violation of this section, is, for reasons of national security, specifically designated by a United States Government Agency for limited or restricted dissemination or distribution"(18 U.S.C. Section 798). Perilously, as explained by Papandrea (2008), Section 798 can reach beyond material that has been classified (pp. 269–270), and the statute does not provide for any requirement that disclosing the information "would pose any harm whatsoever to the United States' national security interests" (p. 270).

In short, Section 798 is potentially the press's formidable Achilles' heel in light of the inclusion of the word *publish,* in addition to "[its] failure to predicate liability upon the finding of a specific intent to harm the United States or benefit a foreign government" (Croner, 2009, p. 770).The most compelling issues in WikiLeaks have their most provocative parallels with and legal provenance in the following case: New York Times Co. v. United States (1971), better known as the Pentagon Papers case.

Daniel Ellsberg was a Pentagon official working with the highest echelon of policy and decision makers in Washington. He expressed his disillusionment with America's foreign policy, particularly with respect to the Vietnam War, secretly photocopying the 7,000 pages, more commonly known as the Pentagon Papers— the "top secret study of U.S. decision making in Vietnam"(Ellsberg, 2002, p. vii.) In 1971, Ellsberg (2002) provided the Pentagon Papers to the New York Times, and the first installment of the Pentagon Papers appeared in its Sunday edition on June 13, 1971 (pp. 386–387). After the third installment, John Mitchell, the Attorney General, asked the New York Times to suspend publication and relinquish its copy of the study (p. 387). With the negative response from the New York Times, the Justice Department made an unprecedented move in filing an injunction in New York's federal district court (Ellsberg, 2002, p. 387). Ellsberg reflects upon the import of this legal posturing:

> For the first time since the Revolution, the presses of an American newspaper were stopped from printing a scheduled story by federal court order....It was the boldest assertion during the cold war that "national security" overrode the constitutional guarantees of the Bill of Rights. (p. 387)

In a 5–4 *certiorari* decision on Friday, June 25, 1971, the Supreme Court decided to review the Second Circuit's decision in favor of the government injunction against the *New York Times*' publication of the Pentagon Papers (Ellsberg, 2002, p. 403). The four dissenting justices, Black, Douglas, Brennan and Marshall, did so, not because they agreed with the injunction, but rather they would have "…immediately lifted all restraints against the *Times* and the *Post*" (Ellsberg, 2002, p. 403). Unprecedentedly, the Supreme Court heard the case on Saturday, June 26, and decided the case on June 30 in a 6–3 decision, with publication resuming July 1 (Ellsberg, 2002, p. 410).

Although Justices Black and Douglas originally voted against *certiorari*, their respective concurring opinions for the majority are impassioned, unmitigated and pellucid. Justice Black states "that every moment's continuance of the injunction against these newspapers amounts to a flagrant, indefensibly and continuing violation of the First Amendment" (*New York Times Co. v. United States*, 1971). He adds the following:

> The press was protected so that it could bare the secrets of government and inform the people. Only a free and unrestrained press can effectively expose deception in government. And paramount among the responsibilities of a free press is the duty to prevent any part of the government from deceiving the people and sending them off to distant lands to die of foreign fever and foreign shot and shell (*New York Times Co. v. United States*, 1971).

Justice Douglas states that the Espionage Act is not applicable to the material which the *New York Times* and the *Washington Post* sought to publish. In Sections 792–799 of Title 18 U.S.C., the word *publish* is mentioned in only three of eight instances {794 (b), 797 and 798}, with the government offering up the word *communicates* as an umbrella term to accommodate the act of publication (*New York Times Co. v. United States*, 1971). Justice Douglas recapitulates the history of Section 793, in that the rejected version was supplanted by the following in the act of Sept. 23, 1950:

> Nothing in this Act shall be construed to authorize, require, or establish military or civilian censorship or in any way to limit or infringe upon freedom of the press or of speech as guaranteed by the Constitution of the United States and no regulation shall be promulgated hereunder having that effect (*New York Times Co. v. United States*, 1971).

Justice Brennan, in his concurring opinion in reaffirming the landmark decision in *Near v. Minnesota* (1931), says, "[that] the First Amendment tolerates absolutely no prior judicial restraints of the press predicated upon surmise or conjecture that untoward consequences may result." (*New York Times Co. v. United States*, 1971). And although allowance is made for overriding prior restraint in times of war, one cannot foretell that the disclosure and publication of information related to war activities (e.g., troop movements, sea transports) would inexorably jeopardize national security interests. In a much more tentative concurring opinion, Justice White concedes that although the disclosure of the documents at the heart of this case is likely to do damage to national security, "I nevertheless agree that the United States has not satisfied the very heavy burden that it must meet to warrant an injunction against publication in these cases..." (*New York Times Co. v. United States*, 1971).

According to Papandrea (2008), in "Lapdogs, Watchdogs, and Scapegoats: The Press and National Security Information," the *per curiam* decision in *New York Times Co. v. United States* (1971) brought the issues of national security and freedom of the press to a face-off (p. 279). With the exception of Justices Black and Douglas, who "protect[ed] the press from criminal laws punishing the publication of classified information as well as prior restraints," (p. 279) on the grounds "that the very purpose of the First Amendment was to protect the ability of the press to 'bare the secrets of government and inform the people,' and that '[s]ecrecy in government is fundamentally anti-democratic, perpetuating bureaucratic errors'" (p. 279), the other justices afforded solid footing to the press predominantly on the issue of prior restraint. The Pentagon Papers irrefutably allowed for the thin edge of the wedge, insofar as emphatic and unequivocal as the Supreme Court was regarding prior restraint as an abrogation of the press's First Amendment rights; the Court was commensurately tenuous regarding the ex post facto criminal action against the press for publishing information about the government (p. 281), and in fact, the government dug in its heels, when jury selection began in the trial of Daniel Ellsberg and Anthony Russo on January 3, 1973 (Arnold, 1973). The allegations included violations of 18 U.S.C.A. Section 371, Section 641 and Section 793. The case against Ellsberg and Russo was dismissed by Judge Byrne on May 11, 1973, in the face of "improper Government conduct shielded so long from public view" (Arnold, 1973).

The WikiLeaks case resonates in some ways similar to and in other ways different from the Pentagon Papers case. In a joint meeting on October 23, 2010, Daniel Ellsberg and Julian Assange repudiated the Obama administration for its dogged pursuits of whistleblowers (Burns & Somaiya, 2010). Ellsberg likened the current administration's "threat to prosecute Mr. Assange to his own treatment under President Richard M. Nixon" (Burns & Somaiya, 2010). In an interview with Amy Goodman of Democracy Now! Ellsberg said the following:

I'm sure, by the way, that if I released the Pentagon Papers today, the same rhetoric and the same calls would be made about me at this time, the same material, same instigations. I would be called not only a traitor, which I was then, which was false and slanderous, but I would be called a terrorist, as a matter of fact. Now, that's the word today for someone who is beyond the pale of any rights, of any rights of citizenship or any human rights, someone who can be just dealt with summarily like that....Assange and Bradley Manning are no more terrorists than I am, and I'm not (Pentagon whistleblower Daniel Ellsberg: Julian Assange is not a terrorist, Dec. 10, 2010).

As Papandrea (2008) states, "Current First Amendment doctrine draws a sharp distinction between the First Amendment rights of those who have received access to classified information as a result of having a position of trust with the government—such as government employees and contractors—and everyone else" (p. 281), with a far different standard applied "to third parties who disclose classified information" (p. 282). So, former and present government employees' First Amendment rights are abrogated predominantly on the grounds of the principle of trust.

Several cases help us understand this relationship and its connection to the WikiLeaks case. In *United States v. Marchetti* (1972), heard in the 4th Circuit (466 F. 2d 1309) and denied *certiorari* by the Supreme Court, the government's injunction against the publication of a book by former CIA employee Victor Marchetti and former State Department employee John D. Marks was upheld. Their book, *The CIA and the Cult of Intelligence* (1974) was eventually published bereft of passages that the government ordered them to delete (Tedford & Herbeck, 2009, p. 228). In this case, the compelling issue guiding the decision was not a First Amendment consideration but rather a contractual one, in which the government does indeed have a justifiable interest in maintaining secrecy (Tedford & Herbeck, 2009, p. 228).

In *Snepp v. United States* (1980), the Supreme Court decided in favor of the government's right to enjoin publication of a work based upon the signing of a secrecy contract. Frank Snepp worked for the Central Intelligence Agency from 1968 to 1976, and upon his departure, he agreed not to divulge any classified information "without the express written consent of the Director of Central Intelligence or his representatives" (Tedford & Herbeck, 2009, p. 229). Snepp published his book without consent, and he was ordered to redirect profits from the book to the government based on the book's earnings (Tedford & Herbeck, p. 229).

In *United States v. Morison* (1988), Samuel Loring Morison's actions set the stage for a case in which Morison, an analyst for the Naval Intelligence Support

Center (NISC), obtained top secret naval photographs and sent them to a British publication, *Jane's Defence Weekly*. For the first time, "a government employee was indicted under the Espionage Act and related statutes for disclosing classified national security information to the press, rather than to a foreign power or its agents" (Papandrea, 2008, p. 296). Morison's failed defense on four counts—two in violation of 18 U.S.C. Section 641 and two in violation of the Espionage Act, 18 U.S.C. Section 793 (d) and (e), circumscribes several areas presaging Bradley Manning's participation in WikiLeaks.

In his concurring opinion in the Morison case, Judge Wilkinson affirms the opinion that, although the interests of national security do not necessarily engender the attenuation of "informed popular debate," insofar as national security can potentially be augmented through the public dissemination of information, nevertheless, as Wilkinson says, "national security is public security, not government security from informed criticism." And in a prescient passage of the opinion, he continues with the idea that, "Confidential diplomatic exchanges are the essence of international relations....When other nations fear that confidences exchanged at the bargaining table will only become embarrassments in the press, our diplomats are left helpless." So the *Dissoi Logoi* is steeped in the promulgation of what Wilkinson calls "public security...compromised in two ways: by attempts to choke off the information needed for democracy to function, and by leaks that imperil the environment of physical security which a functioning democracy requires" (*United States v. Morison*, 1988).

Morison's defense under Sections 793 (d) and (e) was unsuccessful, arguing that his actions were not "classic spying and espionage activity" and that he did not transmit information to an agent of a foreign power, but to a "recognized international news organization located in London...." (*United States v. Morison*, 1988). In short, as recounted by Circuit Judge Russell, Morison's defense is that "he leaked to the press" (*United States v. Morison*, 1988). This argument implies a mutual exclusivity that is facially untenable. What is curious is that Morison furnished the material to a British publication. Additionally, contrary to the Branzburg decision (discussed next), Morison argues that leaks to the press are exempt under Sections 793 (d) and (e).

In *Branzburg v. Hayes* (1972), according to Justice White in writing the majority opinion, the issue turns on "whether requiring newsmen to appear and testify before state or federal grand juries abridges the freedom of speech and press guaranteed by the First Amendment." According to Justice White, notwithstanding the importance of first amendment protection for news gathering, "those cases involve no intrusions upon speech or assembly, no prior restraint or restriction on what the press may publish, and no express or implied command that the press publish what it prefers to withhold." Justice White refuses "to grant newsmen a testimonial privilege that other citizens do not enjoy" (*Branzburg v. Hayes*, 1972). And in spite of this 5–4 opinion, the dissent by Justice Potter Stewart argues for "a qualified privilege of source confidentiality," with the three-part test: relevance, alternative means and compelling interest (Tedford & Herbeck, 2009, p. 245).

And this case is arguably more important when considering which news gatherers and disseminators are (or ought to be) classified as bona fide journalists—with the qualified privilege and First Amendment protection—and which are not. This point acquires new relevance and thinking when looking at Julian Assange and WikiLeaks, gathering and disseminating within the 5th estate modality of communication.

In her article, "Citizen Journalism and the Reporter's Privilege," Papandrea (2007) comments on the volatility of shield laws in light of the changing visage of the reporter's privilege, vis-à-vis First Amendment protection (with the escalating use and significant impact) of the Internet (516). Because the reportorial goal is the dissemination of information, Papandrea says, "the privilege should not be limited to those who are serving as traditional journalists; rather, it should extend to anyone who is contributing to the marketplace of ideas by disseminating information to the public" (p. 519). For Papandrea, "political blogs are some of the most popular and well-known contributors to the public debate" (p. 523), and as evidence, one can argue that online publications like the Huffington Post or Politico rival the best of our traditionally 4th estate news sources. In fact, the mainstream print publications have co-opted many of the strategies used by bloggers with comparable counterparts in their online versions, most notably the *New York Times'* "The Opinionator," introduced in January 2006 (Papandrea, 2007, p. 531). And although in *Zeran v. AOL* (1997) the Internet service providers are not held liable for defamatory content under Section 230 (1) of the Communications Decency Act— "No provider or user of an interactive computer service shall be treated as the publisher or speaker of any information provided by another information content provider" (Carter, Franklin, & Wright, 2008, p. 885)—this exemption does not preclude assigning accountability to the originator of the defamatory message (Carter, Franklin, & Wright, 2008, p. 886). This immunization of service providers was done by Congress in consideration of the interests of more speech against the chilling effect with more attenuated protection (Carter, Franklin, & Wright, 2008, p. 886).

The WikiLeaks incidents pose some interesting and rather thorny dilemmas. Papandrea (2007) states that the reporter's privilege encompasses protection of two types of information: confidential source identity and material attendant to news gathering (pp. 535–536). She argues that W. Mark Felt (infamously known by the soubriquet "Deep Throat") was the famous whistleblower who would not have done what he did without protection as a source (p. 536). Daniel Ellsberg was not an anonymous source, and in fact, he was transparent about his identity as the source of the leak. Bradley Manning ostensibly remained a confidential source for Julian Assange and WikiLeaks, although there exists conflicting evidence. Assange has said that he does not know the identity of the source(s) for the dumped documents that ultimately made their way to the mainstream media, but Manning presumably tells a different version. In one of his chats with Adrian Lamo, discussing the transference of two videos to Wikileaks, he said, "I'm a source, not quite a volunteer," and shortly thereafter, in the same chat, he said, "I mean, I'm a high profile source…and I've developed a relationship with Assange…but I don't know much more than what he tells me, which is very little" (Poulsen & Zetter, 2010). In the *Frontline* interview, "WikiSecrets," Assange made the following statements: "We do not know whether Mr. Manning is our source or not. And of course, if we did know, we are obligated ethically to not reveal it….Our help desk has a completely anonymous chat. It's anonymous to us. The user names are anonymous, and so on" (Gaviria, 2011). Papandrea says that "since September 11, 2001, government employees have played an especially important role in revealing government corruption, illegality and incompetence" (p. 537). Now one could split hairs and say that government workers are obligated differently from military personnel—Felt and Ellsberg being more traditional government employees as opposed to Manning—but Papandrea cultivates a more common bond in saying, "without information from government and military officials, the government's illegal behavior and incompetence may have never come to light" (pp. 537–538). And those federal personnel who do engage in whistleblowing and leaking within the province of national security are particularly vulnerable to the retaliatory tactic, potentially with grave consequences (pp. 542–543). Recall the politicians who believe that Manning's actions warrant a punishment of death. Assange, too, could be accused of conspiracy if the determination were made that he knowingly received information from someone who retrieved the information illegally. Papandrea reinforces the fact that the exception inherent in the Branzburg decision, witnessing or participating in a crime, ought not apply "to individuals who witness the 'crime' of leaking classified or national security information" (pp.

587–588). She argues that leaks of government information, whether classified or not, have become an essential means by which the public learns about government activities (p. 588). She attributes the augmented use of leaking as a means of disseminating information to the inadequacy of the protection for whistleblowers, and although leaks are not by definition in service to the public, many of them are (p. 588). Papandrea even suggests that the standard against prior restraint set in *New York Times Co. v. United States* should apply to reporter's privilege absent the exception circumscribing "a direct and imminent threat to national security or the threat of reasonably certain death or serious bodily harm to another being" (p. 589). And Papandrea reminds us that, in spite of "the government's propensity to exaggerate or manufacture a threat to national security" (p. 589), "in the context of the reporter's privilege, this exception might come into play if a reporter has knowledge of the location and timing of a ticking bomb" (p. 589).

The case of *United States v. Rosen* (2009) might provide the most instructive corollary to the legal issues inherent in WikiLeaks. As discussed in the case summary, Steven Rosen and Keith Weissman, working for the pro-Israel lobbying group AIPAC, were charged with "conspiring to transmit information relating to the national defense to those not entitled to receive it, in violation of 18 U.S.C.S. Section 793 (g)" (*United States v. Rosen*, 2006). As an act of conspiracy prosecuted under the Espionage Act, Lawrence Franklin, their alleged co-conspirator, had top security clearance in the office of the Secretary of the Department of Defense. Rosen and Weissman obtained the information and transmitted it to foreign government officials, media and foreign policy analysts; Rosen was also charged with violating 18 U.S.C. Section 793 (d).

Rosen and Weissman also challenged the application of 18 U.S.C. Section 793 (e) on the grounds that the "statute's indeterminate language failed to provide these defendants with adequate warning that their conduct was proscribed," and that through this proscription, their First Amendment rights of free speech and right to petition were abrogated (*United States v. Rosen*, 2006).

There are several key issues worth delineating here. Defendants argue that Section 793 cannot be applied here because the national defense information (NDI) was transmitted orally. In his opinion, Judge Ellis rejects this argument, saying "[that] the word 'information' is a general term, the plain meaning of which encompasses knowledge derived from both tangible and intangible sources" (*United States v. Rosen*, 2006). Part and parcel of the above argument, defendants contend that the phrase "national defense" is too ambiguous to be clearly defined when that category of information is communicated orally. And because the information was communicated orally, defendants argue that there were no classificatory distinctions (top secret, secret, or confidential), preventing them from knowing whether the transmission of such information would be proscribed. Interestingly, although the "willfulness" requirement "obligates the government to prove that the defendants knew that disclosing the NDI could threaten the nation's security and that it was illegal" (*United States v. Rosen*, 2006), the defendants' actions are still prosecutable irrespective of what Judge Ellis calls "salutary motive" or an "act of patriotism."

The defendants' defense pivoted on the observation that "'leaks' of classified information by non-governmental persons have never been prosecuted under this [793] statute" (*United States v. Rosen*, 2006). Judge Ellis elaborates on the meaning of "leak," which does not preclude its unauthorized transmission but rather the determination as to whether a transmission is contravened—as it was in the case of *United States v. Rosen*.

The secrets that Rosen and Weissman were accused of disclosing reached the level of NDI—which, according to Judge Ellis, are "government secrets the disclosure of which could threaten the security of the nation" (*United States v. Rosen*, 2006). Although Rosen and Weissman argue as well that they were not bound by non-disclosure agreements as government employees would be (e.g., Snepp, Marchetti, & Morison), as mentioned earlier, Judge Ellis found them culpable as violating Section 793 without being in a position of trust within the government.

Assange can be likened to Rosen and Weissman were it not for the former's role as a "citizen journalist" of sorts, as well as an avowed recipient of anonymously leaked information, notwithstanding the disputability of the latter fact. However, Judge Ellis says "that the government can punish those outside of the government for the unauthorized receipt and deliberate retransmission of information relating to the national defense" (*United States v. Rosen*, 2006).Interestingly, toward the end of the opinion in *United States v. Rosen*, Judge Ellis relies on *United States v. Morison*, which states, "The espionage statute has no applicability to the multitude of leaks that pose no conceivable threat to national security, but threaten only to embarrass one or another high government official"(*United States v. Morison*, 1988). Papandrea (2008), in her critique of Judge Ellis's opinion, says the following:

> One clear misstep Judge Ellis made in his application of the strict scrutiny standard is that he required the government to show merely that information is "potentially" harmful. Although this might be the appropriate standard in cases involving the disclosure of government information by government actors, who are obligated contractually and

otherwise to maintain government secrets, it is not an appropriate standard to apply to third parties. The government does not have a compelling interest in silencing information that possesses only potential to harm the national security interests of the United States. Accordingly, the government should be required to prove at a minimum that any such threatened harm is serious, direct, and imminent (p. 298).

# Patriotism and/or Perfidy: Dangerous Heroes in Defiance of the False Dilemma

In recent months, there has been a spate of scholarly articles broaching the First Amendment issues surrounding WikiLeaks. In the wake of, and as a response to, WikiLeaks' activities, the most compelling development has been Congress's introduction of the Securing Human Intelligence and Enforcing Lawful Dissemination Act, known by its acronym, SHIELD (Stone, 2011, p. 105). Although Stone (2011) affords room for the proposed act's constitutionality "[when] applied to a government employee who 'leaks' such classified material, it is plainly unconstitutional as applied to other individuals who might publish or otherwise disseminate such information" (p. 105). Stone also leaves some wiggle room for the act's constitutionality "to those circumstances in which the individual who publicly disseminates classified information knew that the dissemination would create a clear and present danger of grave harm to the nation or its people" (p. 118). In their seminal article, Edgar and Schmidt (1973) raise the issue of the improper classification of classified material as a legitimate "question of fact for the jury" (p. 1065). William Leonard, a government official who directed the Information Security Oversight Office, filed a complaint against officials who, according to Leonard, misclassified a document containing no secrets, singling out a particular e-mail handled by Thomas Drake (Shane, 2011a).

Thomas Drake, a former National Security Agency (N.S.A.) official, was charged with—among other crimes—five counts in violation of 18 U.S.C. Section 793 (e), having taken his concerns about the agency's fiscal mismanagement to a journalist with the *Baltimore Sun* (Shane, 2010a).

Drake was charged with taking classified information to his residence, creating an encrypted e-mail account, as well as shredding documents and deleting them (Shane, 2010). Although the major charges were dropped under the Espionage Act (Shane, 2011a), in June 2011, Drake pleaded guilty "to a misdemeanor of misusing the agency's computer system by providing 'official N.S.A. information' to an unauthorized person, a reporter for the *Baltimore Sun*"

(Shane, 2011b). Like the besieged individuals in whose footsteps he walks, Drake has supporters, including Jesselyn Radack, a lawyer for the Government Accountability Project, who said, "This is a victory for national security whistle-blowers and against corruption inside the intelligence agencies. No public servant should face 35 years in prison for telling the truth" (Shane, 2011b).

There are different dimensions of balancing that must be wrestled with legally and ethically. In the aforementioned admonition that personnel like Drake should not be imprisoned for telling the truth, should that hold for all information to which an individual is privy? Are there compelling reasons for revealing information in light of a perceived greater good, contravening the more fundamental mandate within the organizational structure—in this case, the government—to maintain secrecy?

In her seminal work, *Secrets*, Sissela Bok (1989) says that secrecy is neither inherently good or bad, but rather that, "a degree of concealment or openness accompanies all that human beings do or say" (p. 9). She admonishes us to take an inductive approach through an examination of "particular practices of secrecy, rather than by assuming an initial evaluative stance" (p. 9). So for Bok, the cachet of secrecy is "intentional concealment" (p. 9). The WikiLeaks "publications" traverse a wide array of material, including military (Afghan and Iraq War Logs) and diplomatic (Cablegate). Secrecy, as a practice of the state, is not new, as evidenced by President Woodrow Wilson's introduction of wartime censorship in 1917 through the Committee on Public Information (Bok, 1989, p.172). As Bok says, "Every government has an interest in concealment; every public in greater access to information" (p. 177), and she wonders, "How many leaders have not come into office determined to work for more open government, only to end by fretting over leaks, seeking new, safer ways to classify documents, questioning the loyalty of outspoken subordinates?" (p. 177). And although the passage of the Freedom of Information Act (FOIA) in 1966, and its subsequent fortification in 1974 (Bok, 1989, p. 178), enhanced access to previously unattainable documents, that modality of access would have indubitably been ineffectual in the dissemination of the Pentagon Papers, and more recently, with the WikiLeaks data and that resulting from the activities of Thomas Drake.

When dealing with issues of war, military secrets present dilemmas tantamount to and overlapping with secrets of state. Thomas Jefferson said of the citizenry's right to information about a possible war, "It is their sweat which is to earn all the expenses of the war, and their blood which is to flow in expiation of the causes of it" (Bok, 1989, p. 191). Bok (1989) is mindful of the futility, at times, for opponents of bureaucracy to succeed in mitigating the amount of secrecy through unsuccessful attempts to declassify documents "stamped [as] Top Secret" (p. 197). And military secrecy is oftentimes distinguished from other types of governmental secrecy on the basis of the threat to national security, the reasoning in part, underscoring the cases against Ellsberg, Manning, WikiLeaks and Drake (p. 202). And Bok poses the following questions against the backdrop of the Pentagon Papers case, as they are legitimately posed in the face of subsequent cases discussed here: "What can individuals do who arrive at the conclusion that certain military secrets are detrimental to the national interest and should be made public? Should they take it upon themselves to breach secrecy? What factors should they take into account?" (p. 205). Like Manning and Drake, Ellsberg and Russo were perceived by some as treasonous and by others as self-sacrificing patriots (p. 206), the cohort of detractors arguing that Ellsberg and Russo, in violating administrative and military secrecy, as well as confidentiality, "had thereby contributed to the erosion of respect for the processes of government and of law, and to doubts abroad about the solidity of United States foreign policy" (p. 206).

In the 1919 sedition case, *Abrams v. United States* (1919), Justice Holmes says the following in his dissent:

A patriot might think that we were wasting money on aeroplanes, or making more cannon of a certain kind than we needed, and might advocate curtailment with success, yet even if it turned out that the curtailment hindered and was thought by other minds to have been obviously likely to hinder the United States in the prosecution of the war, no one would hold such conduct a crime.

Holmes does provide a qualification by saying the following:

The power undoubtedly is greater in time of war than in time of peace because war opens dangers that do not exist at other times. But as against dangers peculiar to war, as against others, the principle of the right to free speech is always the same. It is only the present danger of immediate evil or an intent to bring it about that warrants Congress in setting a limit to the expression of opinion where private rights are not concerned.

Ellsberg's dilemma is an exemplary case study, albeit with a twist, applying Kierkegaard's teleological suspension of the ethical. Kierkegaard uses the biblical event of the Akedah—the binding of Isaac by his father Abraham, the latter called upon to sacrifice Isaac to prove his faith in God. In pursuit of the telos—demonstrating his faith in God, Abraham must violate a universal ethical principle (Gordis, 1976, p. 414). As Bok (1989) states, "The conflict of loyalties Ellsberg describes was sharp; it seemed impossible not to betray either his obligations as a civil servant or his conscience and what he believed the public had a right to know" (p. 206). She also questions whether they could avail themselves of any other means in achieving their goal. She suggests that an anonymous leak to either journalists or politicians or speaking about the documents without furnishing the actual documents might have put them in a less ethically dubious position, but the efficacy of either of those approaches is at best indeterminable (p. 209). She does imply a certain virtue in Ellsberg and Russo for having leaked the documents with full self-attribution and with unadulterated understanding of the consequences: "In justifying their choice of civil disobedience they could invoke a principle similar to the one applied by the Nuremberg Tribunal, holding that individuals are at times obligated to disobey government orders that violate international law" (p. 209). Both the whistleblowing and leaking activities have been attributed to Daniel Ellsberg, Bradley Manning, Julian Assange, and Thomas Drake. According to Bok, a whistleblower, "make[s] revelations meant to call attention to negligence, abuses, or dangers that threaten the public interest. They sound an alarm based on their expertise or inside knowledge, often from within the very organization in which they work" (1989, p. 211). The United States Code of Ethics, as applicable to "government servants asks them to 'expose corruption wherever uncovered,' and to 'put loyalty to the highest moral principles and to country above loyalty to persons, party, or Government department'" (pp. 211–212).

Leaking, on the other hand, also discloses information, but it does so in covert fashion and is actually communicated anonymously (Bok, 1989, p. 216). Unlike the whistleblower, whose raison d'être is to expose negligence, abuse and risks, along with the assignation of the determined culpability (Bok, 1989, p. 214), leaking affords transparency to secret information not characterized by any determined "danger, negligence, or abuse" (Bok, 1989, p. 217). But there can be overlap, "when a leak from within does concern misconduct, it is a variant of whistleblowing, undertaken surreptitiously because the revealer cannot or does not want to be known as its source" (Bok, 1989, p. 217).

## Reaching the Goal

When asked about his goals in publishing the cache of WikiLeaks documents, Assange had the following exchange in an interview with Richard Stengel, managing editor of *Time* magazine (*Time's* Julian Assange Interview, 2010):

> RS:     I want to ask you a broader question, about the role of technology and
>         the burgeoning world of social media. How does that affect the goal

you're trying to achieve of more transparent and more open societies? I assume that enables what you're trying to do.

JA:     Let me talk about transparency for a moment. It is not our goal to achieve a more transparent society; it's our goal to achieve a more just society. And most of the times, transparency and openness tends [sic] to lead in that direction....

For Manning, a particular event that occurred within the 7-month period that he spent at the Contingency Operating Station in Iraq catalyzed his feelings about the war, ultimately leading to his decision to mine the databases: "everything started slipping....I saw things differently. I had always questioned the [way] things worked, and investigated to find the truth...but that was a point where I was a part of something, actively involved in something I was completely against" (Leigh & Harding, 2011, pp. 30–31). He said of the decision to put the documents in the public domain, "It's important that it gets out....I feel for some bizarre reason [that] it might actually change something" (Leigh & Harding, 2011, p. 31). Daniel Ellsberg's rationale for leaking the Pentagon papers was conceived in distinct phases:

> For eleven years, from mid-1964 to the end of the war in May 1975, I was, like a great many other Americans, preoccupied with our involvement in Vietnam. In the course of that timeI saw it first as a problem, next as a stalemate, then as amoral and political disaster, a crime. (Ellsberg, 2002, p. vii)

Ellsberg reflects upon the recourse that he had for each phase:

> When I saw the conflict as a problem, I tried to help solve it; when I saw it as a stalemate, to help us extricate ourselves, without harm to other national interests; when I saw it as a crime, to expose and resist it, and to try to end it immediately. (Ellsberg, 2002, p. vii)

If the government pursues conspiracy charges against Assange, given an alleged conspiratorial relationship with Manning, Yale Professor Jack Balkin expounds on the hypothetical scenario in saying, "If you could show that he specifically conspired with a government person to leak the material, that puts him in a different position than if he is the recipient of an anonymous contribution. If he's just providing a portal for information that shows up, he's very much like a journalist" (Savage, 2010). And of course, if Assange, at the helm—presumably of a journalistic enterprise—is accused of publishing material from anonymous sources, bereft of a conspiracy charge, other journalistic enterprises would be more vulnerable in their disclosures of classified documents (Savage, 2010).

And by extension, in furtherance of that slippery slope, how vulnerable is the *New York Times,* for example, to the whim and caprice of a governmental charge, or to the readers of the *New York Times,* or anyone downloading material (Democracy Now!, 2010).

The 1985 case, *United States v. Zehe* (1985), provides for the prosecution of extraterritorial acts of espionage, not only of citizens but of noncitizens as well. In his article, "The Espionage Act and Today's 'High-Tech Terrorist,'" Jamie Hester (2011) says that the Espionage Act applies not only to citizens and noncitizens in a foreign jurisdiction but on the Internet as well (pp. 186–187).

However, as one fine tunes each of his respective motives and concomitant goals, irrespective of the government's rather consistent take on Ellsberg, Manning, Drake and Assange—albeit with variations on the theme—one aspect does emerge, and that is that each is steeped in the ethic of communitarianism. According to Patterson and Wilkins (2010), "communitarianism focuses on the outcome of individual ethical decisions analyzed in light of their potential to impact society. And when applied to journalism, you have a product 'committed to justice, covenant and empowerment. Authentic communities are marked by justice; in strong democracies, courageous talk is mobilized into action'" (p. 14). Although means are important, ends are tantamount to, or possibly paramount to, the means. Individual interests, superseded by more significant interests than the self and self-interest, are certainly the hallmark of the aforementioned leaks and whistleblowers. The ethical dilemma—instantiated in the Kierkegaardian teleological suspension of the ethical—is the untenable role of "serving two masters." In its infamous history, Ellsberg, Manning and Drake have confronted and are still confronting this dilemma squarely, while for Assange and WikiLeaks, the "confrontation" is seemingly more circumventive. On Monday, June 13, 2011, the 40th anniversary of the publication of the Pentagon Papers, the federal government released them in their entirety and in their "original form" (Cooper & Roberts, 2011). In one of his most recent reflections on the declassification of the once classified documents, Ellsberg said, "The reasons are very clearly domestic political reasons, not national security at all. The reasons for the prolonged secrecy are to conceal the fact that so much of the policy making doesn't bear public examination. It's embarrassing or even incriminating" (Cooper & Roberts, 2011). Ellsberg extolled the virtues of military personnel like Manning, saying, "If he did what he's accused of, then he's my hero, because I've been waiting for somebody to do that for 40 years. And no one has" (Cooper & Roberts, 2011). These high-tech leaks compel us to confront fundamental questions regarding the parameters of the journalistic enterprise and its concomitant legal and ethical implications. As Papandrea (2011) states, "It is not possible to draw lines based on the medium of communication, the journalistic background of the publisher, the editing process, the size of the audience, or the methods used to obtain the information" (p. 119). And rather than prematurely rendering a wholesale condemnation of "non-traditional media websites" (Papandrea, p. 124), Papandrea suggests that "we need to begin to recognize that new technology allows non-professionals to play an important role in informing the public (p. 124).

## The Wave of the Future?

Daniel Domscheit-Berg, who worked side by side with Julian Assange, "...perfecting WikiLeaks' technical architecture" (Leigh & Harding, 2011, p. 49), left WikiLeaks on September 15, 2010, and registered the domain name OpenLeaks for his new venture. Domscheit-Berg believes that WikiLeaks' flaws lie in its insistence upon wearing too many hats, so to speak (Domscheit-Berg, 2011, p. 269). WikiLeaks was arguably top-down with Assange at the helm (Domscheit-Berg, 2011, p. 270), and it was centralized, insofar as no separation existed between documents received and documents published (Domscheit-Berg, 2011, p. 271). As lofty as Wikileaks' whistleblowing goal is, OpenLeaks would only serve the receipt function bereft of the publishing function (Domscheit-Berg, 2011, p. 272) "...as a kind of sober, neutral infrastructure" (Domscheit-Berg, 2011, p. 276). To date, Openleaks is inactive.

## The High Dive off the Social Media Platform

Sufian Belhaj, who as Hamadi Kaloutcha provided Arabic and French translations of the Cablegate documents published by WikiLeaks, which he in turn posted on Facebook, wonders whether Mohamed Bouazizi read those translations (Pfeffer, 2012, p. 92). Why is this an important question? Mohamed Bouazizi was the Tunisian fruit vendor who immolated himself on December 17, 2011, as an act of protest—an act that was instrumental in inciting the revolution, toppling not only Tunisia's ruler but other despotic rulers and regimes across the Arab world (Worth, 2011), also known as the Arab Spring. Belhaj's conclusion is that, "[while] WikiLeaks, Twitter and Facebook were the fuel for the revolution, Bouazizi was the spark that ignited it all" (Pfeffer, 2012, p. 93). One can appreciate the metonymic parallels between Belhaj's conclusion and that of the landmark case *Gitlow v. New York* (1925), in which Benjamin Gitlow's printed manifesto was likened in the majority opinion to "a single spark [which] may kindle a fire that, smoldering for a time, may burst into a sweeping and destructive conflagration."

In his essay, "The US Insistence on Internet Freedom Does More Harm Than Good," Clay Shirky (2012) observes that "…since the rise of the Internet in the early 1990s, the world's networked population has grown from the low millions to the low billions (pp. 154–155). Simultaneously, social media have exerted a proportional influence (Shirky, 2012, p. 155), but Shirky poses the following: "How does the ubiquity of social media affect U.S. interests, and how should U.S. policy respond to it?" (p. 155). He also asks, "How much can these entities be expected to support freedom of speech and assembly for their users?" (p. 165). The jury seems to be out on both fronts, although Keller believes that, "the most palpable legacy of the WikiLeaks campaign for transparency is that the U.S. government is more secretive than ever" (Keller, 2012).

What can be said irrefragably is that Marshall McLuhan's dictum, "the medium is the message," is alive and well—the ramifications of which—only politics, culture, technology and time will tell. The flames will have been doused and the air will have cleared, but in the end, WikiLeaks' cyberpyrotechnics might very well have been a pyrrhic victory.

## References

*Abrams v. United States*, 250 U.S. 616 (1919).

Arnold, M. (1973, January 4). Defense sees constitutional test as Ellsberg-Russo trial starts. *New York Times*. Retrieved from http://www.nytimes.com

Arnold, M. (1973, May 12). Pentagon Papers charges are dismissed; Judge Byrne frees Ellsberg and Russo, assails 'improper government conduct.' *New York Times*. Retrieved from http://www.nytimes.com

Batty, D. (2010, August 27). WikiLeaks war logs posting 'will lead to free speech ruling.' *Guardian*. Retrieved from http://www.guardian.co.uk

Bok, S. (1989). *Secrets*. New York: Vintage Books.

*Branzburg v. Hayes*, 408 U.S. 665 (1972).

Bryfonski, D. (Ed.). (2012). *The global impact of social media*. Detroit, MI: Greenhaven.

Burns, J., & Somaiya, R. (2010, October23). WikiLeaks founder gets support in rebuking U.S. on whistle-blowers. *New York Times*. Retrieved from http://www.nytimes.com

Carter, T., Franklin, M., & Wright, J. (2008). *The first Amendment and the fifth estate*. New York: Foundation Press.

Choate, M. (2008). *Professional Wikis*. Indianapolis, IN: Wiley.

Clinton, Hillary, (2012, Nov. 29). U.S. 'deeply regrets' embarrassment of WikiLeaks documents. PBS NewsHour.Retrieved from http://www.youtube.com

Cooper, M., & Roberts, S. (2011, June 7). After 40 years, the complete Pentagon Papers. *New York Times*. Retrieved from http://www.nytimes.com

Croner, A. (2009). A snake in the grass?: Section 798 of the Espionage Act and its constitutionality as applied to the press. *The George Washington Law Review*, 77, 766–798. Retrieved from http://www.lexisnexis.com

Davies, N., & Leigh, D. (2010, July 25). Afghanistan war logs: Massive leak of secret files exposes truth of occupation. *The Guardian*. Retrieved from http://www.guardian.co.uk

de Nies, Y., & Miller, S. (2010, July 27). Obama on WikiLeaks: Documents don't reveal any issues that haven't already informed our public debate [Web log post]. Retrieved from http://blogs.abcnews.com

Department of Defense. (2010, July 5).Charge Sheet: Manning, Bradley E.

Department of Defense. (2011, March 1). Charge Sheet: Manning, Bradley E.

Domscheit-Berg, D. (2011). *Inside WikiLeaks* (Jefferson Chase, Trans.). New York: Crown Publishers.

Edgar, H., & Schmidt, B. C. (1973). The Espionage Act and publication of defense information. *Columbia Law Review*, 7(5), 929–1087. Retrieved from http://www.jstor.org.ezproxy.ic.edu/stable/view/1121711

Ellsberg, D. (2010, Nov. 30). Interview by Richard Stengel. [Skype].Transcript published Dec. 1, 2010. Retrieved fromhttp://www.time.com

Ellsberg, D. (2010, Dec. 10). Radio Interview by Amy Goodman. Pentagon whistleblower Daniel Ellsberg: Julian Assange is not a terrorist, *Democracy Now!* http://www.democracynow.org/2010/12/10/whistleblower_daniel_ellsberg_julian_assange_is

Ellsberg, D. (2002). *Secrets: A memoir of Vietnam and the Pentagon papers*. New York: Penguin Books.

Gaviria, M. (Writer & Director). (2011). *WikiSecrets* [Television documentary]. On *Frontline*. New York: PBS.

*Gitlow v. New York*, 268 U.S. 652 (1925).

Gladwell, M. (2012). Social media fail to incite true activism. In D. Bryfonski (Ed.),*The global impact of social media* (pp. 95–105). Detroit, MI: Greenhaven Press.

Gordis, R. (1976). The faith of Abraham: A note on Kierkegaard's "teleological suspension of the ethical." *Judaism*, 25, 414–419. Retrieved from http://www.ajcongress.org

Hester, J. (2011, June). The Espionage Act and today's high-tech terrorist. *North Carolina Journal of Law & Technology, Online Edition*.Retrieved from http://www.lexisnexis.com

Keller, B. (2011, January 26). Dealing with Julian Assange and the secrets he spilled. *New York Times*. Retrieved from http://www.nytimes.com.

Keller, B. (2012, February 19). WikiLeaks, a postscript. *New York Times*. Retrieved from http://www.nytimes.com

Khatchadourian, R. (2010, June 7). No secrets: Julian Assange's mission for total transparency. *New Yorker*. Retrieved from http://www.newyorker.com

Leigh, D. (2010, Nov. 28). How 250,000 US embassy cables were leaked. *The Guardian*. Retrieved from http://www.guardian.co.uk

Leigh, D. (2010, 22 Oct.). Iraq war logs: An introduction. *Guardian*. Retrieved from http://www.guardian.co.uk

Leigh, D., & Harding, L. (2011). *Wikileaks: Inside Julian Assange's war on secrecy*. New York: Public Affairs.

Loader, B., & Mercea, D. (2012). Networking democracy? Social media innovations in participatory politics. In B. Loader & D. Mercea (Eds.), Social media and democracy: Innovations in participatory politics (pp. 1–10). London: Routledge.

Mainwaring, S. (2012). Social media empower people. In D. Bryfonski (Ed.), The global impact of social media (pp. 87–90). Detroit, MI: Greenhaven Press.

Marchetti, V., & Marks, J. (1974). The CIA and the cult of Intelligence. New York: Knopf.

Morozov, E. (2012). The role of social media in Iran was exaggerated. In D. Bryfonski (Ed.),The global impact of social media (pp. 106–115). Detroit, MI: Greenhaven Press.

Near v. Minnesota, 283 U.S. 697 (1931).

New York Times Co. v. United States, 403 U.S. 713 (1971).

Papandrea, M. (2007). Citizen journalism and the reporter's privilege. Minnesota Law Review, 91,516–591. Retrieved from http://www.lexisnexis.com

Papandrea, M. (2008). Lapdogs, watchdogs, and scapegoats: The press and national security information. Indiana Law Journal, 83, 234–305. Retrieved from http://www.lexisnexis.com

Papandrea, M. (2011, June 15). The publication of national security information in the digital age. Journal of National Security Law & Policy, 5, 119–130. Retrieved from http://www.lexisnexis.com

Patterson, P., & Wilkins, L. (2010). Media ethics: Issues and cases. New York: McGraw-Hill.

Pentagon Whistle Blower Daniel Ellsberg: Julian Assange is not a terrorist, Huff Post Politics, Democracy Now, http://www.huffingtonpost.com/democracy-now/pentagon-whistleblower-da_b_795045.html

Pfeffer, A. (2012). The Tunisian revolt was the world's first Facebook revolution. In D. Bryfonski (Ed.), The global impact of social media (pp. 91–105). Detroit, MI: Greenhaven Press.

Poulsen, K., & Zetter, K. (2010, June 10). 'I can't believe what I'm confessing to you': The WikiLeaks chats. Wired. Retrieved from http://www.wired.com

Savage, C. (2010, December 7). U.S. Prosecutors study WikiLeaks prosecution. New York Times. Retrieved from http://www.nytimes.com

Shane, S. (2010, June 11). Obama takes a hard line against leaks to press. New York Times. Retrieved from http://www.nytimes.com

Shane, S. (2011a, August 1). Complaint seeks punishment for classification of documents. New York Times. Retrieved from http://www.nytimes.com

Shane, S. (2011b, June 9). Ex-N.S.A. aide gains plea deal in leak case; setback to U.S. New York Times. Retrieved from http://www.nytimes.com

Sharma, P. (2012). Social media did not cause the Egyptian revolution. In D. Bryfonski (Ed.),The global impact of social media (pp. 116–118). Detroit, MI: Greenhaven Press.

Shirky, C. (2012). The US insistence on internet freedom does more harm than good. In D. Bryfonski (Ed.),The global impact of social media (pp. 154–166). Detroit, MI: Greenhaven Press.

Snepp v. United States, 444 U.S. 507 (1980).

Stelter, B., & Cohen, N. (2011 April 26). In WikiLeaks' growth, some control is lost. New York Times. Retrieved fromhttp://www.nytimes.com

Stone, G. (2011, June 15). WikiLeaks, the proposed SHIELD Act, and the First Amendment. Journal of National Security Law & Policy, 5, 105–118. Retrieved from http://www.lexisnexis.com

Tedford, T., & Herbeck, D. (2009). Freedom of speech in the United States. State College, PA: Strata.

Time's Julian Assange interview (2010, Dec. 1). *Time Magazine*. http://www.time.com/time/world/article/0,8599,2034040,00.html

Title 18 U.S.C. §§793–798.

*United States v. Marchetti*, 466 F2d. 1309 (4th Cir. 1972).

*United States v. Morison*, 844 F.2d 1057 (1988).

*United States v. Rosen*, 557 F. 3d 192 (E.D. Va., 2009).

*United States v. Rosen*, 445 F. Supp. 2d 602 (E.D. Va. 2006).

*United States v. Zehe*, 601 F. Supp. 196 (U.S. Dist. Mass., 1985).

Wikipedia. (n.d.). FAQ/Overview. How big is Wikipedia? Retrieved from http://en.wikipedia.org/wiki/Wikipedia:FAQ/Overview

Worth, R. (2011, January 21). How a single match can ignite a revolution. *New York Times*. Retrieved from http://www.nytimes.com

*Zeran v. America Online*, 129 F.3d. 327 (4th Cir. 1997).

# · 7 ·

# Revisiting the Right "To Be Let Alone" in the Age of Social Media

WARREN SANDMANN

The right of privacy, conceding it to exist, is a purely personal one, that is it is a right of each individual to be let alone, or not to be dragged into publicity. (Cooley & Lewis, 1907, p. 195)

You have zero privacy anyway. Get over it. (Sprenger, 1999)

These two quotations provide the range for this chapter: What is the role of the private in the age of the social? The first quotation is taken from one of the classic works on the law of torts: Thomas Cooley's *A Treatise on the Law of Torts or the Wrongs Which Arise Independently of Contract* (Cooley & Lewis, 1907). The second quotation is attributed to a former executive (Scott McNealy) of a technology company, Sun Microsystems (a company that no longer truly exists, having been purchased by a larger technology company, Oracle—and the reader can infer any meaning he or she wants into the idea of an oracle in charge of our information).

To answer this question, we begin by revisiting Warren and Brandeis's 1890 essay, "The Right to Privacy." Following the review, we look at the evolution of the principles described by Warren and Brandeis, making the argument that the concept of privacy developed by them is primarily a product not just of a time period but of a social class, is not relevant to contemporary culture, and does not fit with the technology of social media. Following this, we discuss the place and function of privacy in a socially networked world. In closing, we return to Warren and Brandeis by noting the continuing relationship between culture and law as it relates to privacy.

# Why a Need for Privacy?

What drove Warren and Brandeis to write their essay on privacy? The answer to this question foreshadows one of the later arguments made, and that answer is, because Warren and Brandeis believed that the social mores and customs of the late 19th century United States were diminishing the quality of life for people like them. The introductory paragraphs of their essay read more as a general complaint by the gentry than a rationale for a legal principle. This is an essay grounded in morality and disapprobation. Here are some examples:

> "Instantaneous photographs and newspaper enterprise have invaded the sacred precincts of private and domestic life…" (Warren & Brandeis, 1890). The choice of the term *sacred* is deliberate and evocative. To Warren and Brandeis, there is a clear distinction between the public and the private, between what can be known, shared, and discussed by all, and what should always and forever remain behind closed doors. More importantly, the private is not only protected, but "sacred," a place separate from the secular world, cloistered, vulnerable to profanity, and valued above any public space.

- "[T]he evil of invasion of privacy by the newspapers, long keenly felt, has been but recently discussed by an able writer" (Warren & Brandeis, 1890). Granted, this was not always an era of fair and objective journalism, with partisan newspapers tied to political parties and very interested in supporting the views of their patrons (Smythe, 2003). Nonetheless, Warren and Brandeis signal their tone of moral outrage and their place in the holy fight for privacy through their word choice: Once again, privacy is the sacred place distinct from the lower, less genteel, and potentially degrading public sphere.
- The press is overstepping in every direction the obvious bounds of propriety and of decency. Gossip is no longer the resource of the idle and of the vicious, but has become a trade, which is pursued with industry as well as effrontery. To satisfy a prurient taste the details of sexual relations are spread broadcast in the columns of the daily papers. To occupy the indolent, column upon column is filled with idle gossip, which can only be procured by intrusion upon the domestic circle. (Warren & Brandeis, 1890)

The argument made here is far more in the sphere of morality than that of legality. The reason for this becomes clearer as Warren and Brandeis advance their argument. What they wish for is to create a general principle for privacy that is distinct from the more materialistic law of property rights, or even of libel, slander, and defamation. They want a concept of privacy that is grounded in the moral superiority of the private. There is a clear difference in their thinking between the sacred private places and the coarse public places, between the "necessary...retreat from the world...under the refining influence of culture" of the private and the "mental pain and distress" of the public (Warren & Brandeis, 1890). The moral nature of this argument is seen most clearly in this last excerpt, as the language presented leaves no doubt as to the concern of the authors:

> Even gossip apparently harmless, when widely and persistently circulated, is potent for evil. It both belittles and perverts. It belittles by inverting the relative importance of things, thus dwarfing the thoughts and aspirations of a people. When personal gossip attains the dignity of print, and crowds the space available for matters of real interest to the community, what wonder that the ignorant and thoughtless mistake its relative importance. Easy of comprehension, appealing to that weak side of human nature which is never wholly cast down by the misfortunes and frailties of our neighbors, no one can be surprised that it usurps the place of interest in brains capable of other things. Triviality destroys at once robustness of thought and delicacy of feeling. No enthusiasm can flourish, no generous impulse can survive under its blighting influence. (Warren & Brandeis, 1890)

The need for privacy, the protection of privacy, then, is not just for the protection of the individual—it is for the moral good of society as a whole. When the private becomes public, all are diminished, both individually and as a society.

## What Does Privacy Mean?

Warren and Brandeis note that the law already recognizes many areas where some sort of "private" information is implicated in and protected by legal remedy. Slander, libel, and defamation: These three address issues of privacy, but they are not necessarily protective of privacy so much as they are protective of personal reputation. As such, Warren and Brandeis argue, they really are more akin to infringements of property rights—slander, libel, and defamation "steal" the good name of a person, take away from the reputation of a person. And while what is slandered, libeled, or defamed may include the private, these laws and the remedies available do not really address the core of the private.

Copyright, similarly, may seem to deal with privacy, and the remedies available there may also seem to offer some additional relief. Not so, argue Warren and Brandeis. Unless compelled to do so in court, individuals retain the complete right to privacy concerning their thoughts and the manner in which they would express those thoughts. Once the decision is made to express those thoughts, however, copyright protects only a property interest in the publication of ideas and images; so, once published, the author of these ideas and images retains a property right but voluntarily gives up a right to privacy.

It is in this idea of "voluntarily" giving up something that we begin to see what Warren and Brandeis mean by privacy. Privacy is the ability of a person to withhold from the public that which a person wishes to withhold. To be private is to maintain control over your ideas, over that information and those images that are at the core of who you are. Privacy is the ability to prohibit photographic images of you to be circulated without your consent and control; privacy is the ability to keep out of the newspaper stories of what you have done. Privacy is, "the more general right to the immunity of the person,—the right to one's personality" (Warren & Brandeis, 1890).

While the purpose of this chapter is not to define privacy (or even wade into the very deep pool that is the concept of privacy), it is important to briefly discuss some of the ways that American society, and American (and Western to an extent) law, have conceived of privacy.

Ken Gormley, in a 1992 essay, offers both a definitional typology and a legal categorization of privacy. Gormley states that there are four ways to define privacy: First, privacy can be defined as "an expression of one's *personality* or *personhood*, focusing on the right of the individual to define his or her essence as a human being" (p. 1332). The second definitional focus for privacy rests in persona autonomy, "the moral freedom of the individual to engage in his or her own thoughts, actions and decisions" (p. 1332). A third definition focuses on control and ownership of information, the ability of the individual to maintain control and authority over what information is made public, when, and to whom (p. 1333). The fourth and final definitional category is actually less a distinct category but more a conglomeration of other theories and definitions that do not cleanly fit the first three categories—privacy defined as a series of concepts and components lacking a clear, central definitional core (p. 1333).

In addition to these definitional elements, Gormley (1992) offers a useful analytical tool for understanding how law has dealt with the concept of privacy. Gormley identifies what he calls "five dominant species of legal reasoning"

(Gormley, p. 1340) starting with and building from Warren and Brandeis's conception:

- ⅄ Tort Privacy
- ⅄ Fourth Amendment Privacy
- ⅄ First Amendment Privacy
- ⅄ Fundamental Decision Privacy
- ⅄ State Constitutional Privacy

## Tort Privacy

The introductory section of this chapter includes a summary of Warren and Brandeis's essay and its concepts, so there is little need to rehash these ideas here. What Gormley (1992) notes, however, is that Warren and Brandeis were essentially trying to bring the law into a sphere where law was largely absent, but where there was, in Gormley's words, a "deeper instinct in common law" (p. 1352). As discussed earlier, Warren and Brandeis's essay sought to develop a way for the law to address what they saw as a necessary need for the individual—a legal way to keep private that which should be kept private. While Warren and Brandeis claim that the concept they developed in their essay was grounded in legal history, and was little more than a logical and legal conclusion of what courts were finding, Gormley notes that if one looks closely at the cases used by Warren and Brandeis, what is actually found is "the fact that a right to privacy clearly did *not* exist...in the year 1890" (p. 1347). Instead, "the truth was that there were shreds and ribbons of privacy adorning the jurisprudence of England and America, but nothing big enough to wrap up and place in a package," such as the type of package Warren and Brandeis were developing (pp. 1347–1348).

## Fourth Amendment Privacy

At its heart, this element of legal privacy is a variation of the concept that "a man's house is his castle" (Paxton's Case, 1761). Gormley quotes William Pitt the Elder:

> The poorest man in his cottage may bid defiance to all the forces of the Crown. It may be frail—its roof may shake—the wind may blow through it—the storm may enter, the rain may enter—but the King of England cannot enter—all his forces dare not cross the threshold of the ruined tenement! (quoted in *Frank v. Maryland, 1959*)

This conception of privacy is, as noted in the heading of this section, most clearly evident in the prohibition against unreasonable and illegal search and seizures codified in the Fourth Amendment and highlighted legally by Brandeis in his dissent in *Olmstead v. United States* (1928), arguing against the ability of the government to engage in phone wiretapping.

In addition to the concept of privacy relating to search and seizure, what is also interesting about this conception of privacy is its connection with technology. *Olmstead v. United States*, after all, was about the use of a new technology that seemed to be outside the traditional definition of "search and seizure." After all, how could simply using technology to hear what others are saying be seizure? This element is explored more fully later in this chapter when we look at social networking.

## First Amendment Privacy

At first glance, it may appear (and even at second glance) that the First Amendment is more antithetical to privacy than it is supportive of privacy. Gormley's (1992) explanation seems accurate: Privacy as related to the First Amendment is not so much something that can be found laid out in the words and meaning of the First Amendment; privacy under the First Amendment is the mirror image, the counterweight, of the First Amendment right to be expressive (p. 1375). If the First Amendment allows expression, then there must be a balance that provides a means to restrict, in some fashion, that expression. We have the right to speak on (almost) any topic at (almost) any place in (almost) any fashion. Privacy under the First Amendment resides in the "almost."

## Fundamental Decision Privacy

This concept of privacy is most clearly enunciated, Gormley (1992) claims (correctly) in two cases: *Griswold v. Connecticut* (1965) and *Roe v. Wade* (1973). *Griswold v. Connecticut* found a majority of justices able to create a fundamental right to at least a certain form of privacy "broad enough to protect the ability of married couples to decide what to do in the privacy of their marital bedrooms" (p. 1392). *Roe v. Wade* took this concept one step further and grounded a woman's right to choose to abort in the right to remain private. Where this concept differs from the earlier ones (though not so greatly, as some would argue) is that it appears to be less grounded in legal theory and even more grounded in changing social mores and technology. However, and Gormley makes this argument in more detail, this concept of privacy is not really all that distant and distinct from the other categories.

Briefly, as discussed earlier, Warren and Brandeis's tort theory was more a construction of social and technological changes than a logical progression of the law. Additionally, *Griswold v. Connecticut* and *Roe v. Wade* both have legal genealogies that are at least as sound as the development of a tort theory of privacy. In retrospect, Justice Douglas could have perhaps chosen a better term than *penumbras* to ground this concept of privacy, but the connections between already established definitions and legal constructs of privacy as that announced in *Griswold v. Connecticut* is not that great a leap of logic or law.

## State Constitutional Privacy

While perhaps on the surface not clearly distinct enough to even merit a category of its own, Gormley (1992) argues that there are distinguishing characteristics that lead to the need for this fifth category. First, many of the state constitutions provide stronger and broader protections for privacy than does federal law; second, some state constitutions have provided for "entirely novel packets of privacy" (p. 1422).

So we have these definitions and categories. So what? What does this mean for our understanding of privacy in a socially networked world? There are two short answers, which are developed more fully in the next section. First, how we conceive of privacy, personally, socially, and legally, is always shaped by how we have conceived of it in the past. Second, social and technological changes precede legal changes—law is (almost) always behind society and technology. So let us proceed to the next section and look at what privacy may mean in a socially networked world.

# Privacy in a Socially Networked World

First, a little more definition: What do we mean by socially networked world? The theoretical perspectives of Erving Goffman and Joshua Meyrowitz provide a good grounding. Goffman (1959) provides us with the concept of the self as performer.

> [A] selected display of behavior that cannot go on continuously and which must, to some extent, consciously or unconsciously, be planned and rehearsed. And, just as in a play, the stage must be properly set, the actors must carefully control their actions,

and the roles in one drama must be kept separate from roles in other dramas. (as cited in Meyrowitz, 1986, p. 29)

Goffman's perspective creates, just as in a theater, a front stage and a back stage—public, performance spaces and private, rehearsal spaces. But these spaces are more than just public and private, performance and rehearsal. In both places, performance takes place—but the performance is different because of the "set" and the "audience." Both the front and back stage are sets, and both have audiences—they just have different rules and different expectations.

Meyrowitz takes this concept one step further with his emphasis on technology and electronic media. In Goffman's initial depiction of front and back stage, there remains an element of control by the individual—a choice to be front or back. There also remains an element of stasis—that front stage and back stage are physical and stable "places," that they remain separate and distinct. Electronic media, and especially social media, change both of these elements.

By severing the traditional link between physical location and social situation, for example, electronic media may begin to blur previously distinct group identities by allowing people to 'escape' informationally from place-defined groups and by permitting outsiders to 'invade' many groups' territories without ever entering them. (Meyrowitz, 1986, p. 57)

So what is a socially networked world? It is a place where distinctions between front stage and back stage "performances" are less distinct—if they are distinct at all. It is a place where rehearsal and performance are almost one and the same. It is a place where choice is not really choice—to be in the socially networked world is, by definition, to choose to be front stage all the time. It is a place where the only stasis is that the individual is always on stage. It is, in Meyrowitz's words, a place with no sense of place—or even more accurately, a place that is all places. The Benthamite concept of the Panopticon comes to mind, with one exception (Bentham, 1995). The Panopticon was a place where those being observed could be observed all the time—but without their awareness. The socially networked world operates on the knowledge (or at least tacit—and maybe *passive* is a better term—understanding) that behavior is being observed— and that being observed is the point of the socially networked world.

Sloop and Gunn (2010) offer what I think is a concise understanding of the socially networked world: "What happens when we no longer fear surveillance but actually experience the gaze as a form of pleasurable or compulsory publicity? Or worse, what happens when we are invited to forget the machineries of power?" (p. 290).

What, then, is private (or at least what is the conception of privacy) in the socially networked world? We look at two elements: choice (or lack of) for individual control and ownership of a person's history.

Read the rules for privacy settings on Facebook. First, that will already place you in a select group. Second, choosing to follow those rules and settings is, for the most part, antithetical to the concept of social networking.

Debatin, Lovejoy, Horn, and Hughes (2009) studied the interaction between knowledge of Facebook privacy settings and actual behavior. People using Facebook claim to be well aware of the dangers of loss of privacy and of making private information available to the public. Despite this stated awareness, users still upload significant amounts of private data, seemingly unconcerned about the dangers of loss of privacy. The authors of the study attribute this discrepancy to how Facebook is used. Those making all this data public do so because they like what Facebook has to offer: They can share information with lots of people. Additionally, these users believe that loss of privacy is something that happens to others, not to themselves. The authors do note that when users have bad experiences with loss of data, they are more careful and do then change their privacy settings—but only when they directly experience this— there is no change from just hearing about this happening to others.

Many have argued that the providers of the services, Facebook specifically, should do more to educate users about these choices in what (and how much) to share. The companies, accurately but perhaps not realistically, note that they always have educated users about these choices—and have continued to evolve their privacy practices (Pachel, 2011). Given these concerns, and the lackluster response by companies such as Facebook, Google, and Amazon, the Federal Trade Commission initiated investigations of the privacy polices of these (and other) companies. The result is that now the government is watching as well (Arora, 2012).

Will this change anything? Pachal (2011) argues, and I think convincingly, that the major change that will come from this oversight is that these companies will work harder to develop and implement more and more clear "opt-in" policies on privacy. On the face, this seems a good thing—more informed decisions, correct? However, as Debatin and coauthors note (2009), knowledge about privacy does not seem to be the problem; caring enough to do something about privacy is more of a problem for most users. Additionally, Pachal also notes that what these companies will really focus on is more deeply embedding those who opt-in into the panoply of interlinked relationships. Once users opt-in, they will be more likely to justify their decision by taking advantage of the connections—those individuals will become superusers of the services.

Privacy is a negotiation: To gain the benefits and advantages of a social net-work—in other words, to use a social network as it is designed to be used—individuals must negotiate the balance between privacy and function. Choosing individual control negates a social network; choosing to take part in a social network means that you give up individual control. The question is no longer whether a person maintains control; the question is how much is a person willing to give up in order to be part of the network.

Some may argue that this choice is no different than any choice to establish a relationship with another person. Basic interpersonal communication theory takes as a given that in order to develop a relationship with another, a person has to give up control of at least some of his or her individual information. The difference here is one of scope. Interpersonal relationships, or even group relationships, are between limited and, for the most part, selected individuals. Choosing to take part in a socially networked world eliminates the selection of individuals with whom a person wishes to share and broadens the scope far beyond the individual or group. As Ardia, (2010) notes, "the Internet connects us to hundreds of millions of people. Our existing notions of how to establish trust and maintain social ties do not always translate to this networked world" (p. 263).

In addition to this (false) choice of individual control, ownership of personal information, especially the ability to maintain ownership of self-identity, is altered in a socially networked world. Once a person has "chosen" to take part in the socially networked world, not only the immediate but also the long-term loss of control and ownership shifts.

In an individual relationship, I can share information about myself to another person or to a group of people. That information is, in a sense, out of my control. Additionally, the individual or individuals with whom I have shared that information may also share it with others—I have little control over that.

But my loss of ownership over that information is still limited in scope—it rests with a small group of people, and even as it is shared, it is diluted and altered and made more difficult to connect back to me. A socially networked world changes that equation. Ardia (2010) notes the following:

> Global communication networks such as the Internet have made reputation more enduring and yet more ephemeral. Reputation is more enduring because information about us, whether good or bad, can exist—and be easily retrievable—forever. Powerful search engines scour and index photos, videos, and text. Semantic connections link previously disparate pieces of information to individuals and to each other. In the past, much personal information was publicly inaccessible because of practical impediments to its access. The Internet is largely eliminating these impediments. (p. 262)

Information I share on a social network is information to which I forever cede ownership and control. I can add to the information, I can try to bury old information with new information, I can hire some company that claims it will remove all the old information about me—but the information remains and it remains accessible in a fashion in which information has never before been accessible.

## Privacy, Culture and Technology

In reviewing Warren and Brandeis's original article, we read the argument that their development of privacy was largely a function of social mores and customs, a response to changes in technology, and a "distaste" for a loss of control—that primarily, this was a moral argument, an argument about what "ought" to be allowed, and additionally that there was perhaps more than a whiff of class-based thinking involved. This argument seems sound, and I believe it continues to play out in our discussion of social networks and privacy.

Writers such as Robert Putnam have made headlines bemoaning the loss of community and the shattering of traditional social networks. New digital networks allow for different connections, different definitions of community, and different definitions of friendship and relationship. Technological changes have made it possible (and some would argue have also driven these social changes) for people to develop and maintain relationships that are no longer place-bound. These new networks have meant changes in how we see ourselves as individuals, how we think about our relationships with others, and how we share information.

The socially networked world is an information world. Information about the individuals who populate this world is the *currency* (and this term is used deliberately) that makes the socially networked world possible and attractive to all users. In order to function in this world, what we have thought of as "private" must change. I am not arguing whether this is a good or bad change. What I do argue in this essay is that our traditional definitions and understanding of privacy cannot continue unchanged in a socially networked world.

This chapter started with two quotations:

1. "The right of privacy, conceding it to exist, is a purely personal one, that is it is a right of each individual to be let alone, or not to be dragged into publicity" (Cooley & Lewis, 1907, p. 195).
2. "You have zero privacy anyway. Get over it" (Sprenger, 1999).

The individual Cooley and Lewis speak of in the first quotation is an individual who cannot exist in a socially networked world, a world where the currency is always available publicly. As a society, in our (current at least) embrace of social media, we appear to have given up at least a good portion of our belief in "the right to be let alone." While we may not be ready to adhere completely to McNealy's belief that there is no privacy anyway, the negotiation between these two positions has shifted significantly in the socially networked world.

## References

Ardia, D. (2010). Reputation in a networked world: Revisiting the social foundations of defamation law. *45 Harv. C.R.-C.L. L. Rev.* 261.

Arora, N. (2012). FTC lays out rules to fight privacy invasions from Apple, Google, Facebook. Retrieved from http://www.forbes.com/sites/greatspeculations/2012/03/28/ftc-lays-out-rules-to-fight-privacy-invasions-from-apple-google-facebook/

Bentham, J. (1995). Panopticon (Preface). In M. Bozovic (Ed.), *The Panopticon writings* (pp. 29–95). London: Verso.

Cooley, T.M., & Lewis, J. (Ed.). (1907). *A treatise of the law of torts or the wrongs which arise independently of contract.* Chicago: Callaghan and Company.

Debatin, B., Lovejoy, J. P., Horn, A., & Hughes, B. (2009). Facebook and online privacy: Attitudes, behaviors, and unintended consequences. *Journal of Computer-Mediated Communication, 15,* 83–108.

*Frank v. Maryland,* 359 U.S. 360 (1959).

Goffman, E. (1959).*The presentation of self in everyday life.* New York: Doubleday.

Gormley, K. (1992). One hundred years of privacy. *1992 Wis L. Rev.* 1335.

*Griswold v. Connecticut,* 381 U.S. 479 (1965).

Meyrowitz, J. (1986). *No sense of place: The impact of electronic media on social behavior.* New York: Oxford University Press.

*Olmstead v. United States,* 277 U.S. 438 (1928).

Pachel, P. (2011). What the Facebook FTC settlement means for social media. Retrieved from http://mashable.com/2011/11/29/facebook-ftc-settlement-2/

Paxton's Case, Superior Ct. 1761, *reprinted in* Quincy's Mass. Rep. 1761–1762, 51.

*Roe v. Wade,* 410 U.S. 113 (1973).

Sloop, J., & Gunn, J. (2010). Status control: An admonition concerning the publicized privacy of social networking. *The Communication Review, 13*(4), 289–308.

Smythe, T. C. (2003).*The gilded age press: 1865–1900.* Westport, CT: Praeger.

Sprenger, P. (1999, January 26). Sun on privacy: 'Get over it.' Retrieved from http://www.wired.com/politics/law/news/1999/01/17538

Warren, S., & Brandeis, L. (1890).The right to privacy. *Harvard Law Review, 4*(5). Retrieved from http://groups.csail.mit.edu/mac/classes/6.805/articles/privacy/Privacy_brand_warr2.html

## · 8 ·

# Transparency, Misrepresentation, and Social Media

DOUGLAS C. STRAHLER & THOMAS R. FLYNN

"Pay no attention to that man behind the curtain."

Chances are at some point in your life, you've seen *The Wizard of Oz*, a staple of pop culture that takes us on a journey with Dorothy and her new friends to find the Wizard. When Dorothy first meets the Wizard, he appears as a large, disembodied head with a deep intimidating voice, surrounded in smoke and flames. Not until they later return to the Emerald City do they learn he is a fraud, when Toto pulls open a curtain to reveal an ordinary man using technology to create an illusion.

Over the years, scholars have examined the movie's symbolism and applied political interpretations to the characters of the movie. Harmon (2009), for example, claims that Toto represents the average American, and in revealing the fraud of the Wizard, demonstrates the story's faith in democratic self-government. The average citizen can see things as they are. Although the movie can be seen as a commentary on the political environment of the late 1800s, an analogy can be drawn to current issues and concerns in contemporary politics and technology. Toto's pulling back the curtain can be seen to represent today's average citizens and how technology has given them the power to create greater transparency in society by pulling the metaphorical curtain away to expose the thoughts and actions of individuals, businesses, corporations, and governments. Furthermore, just as technology was used in the Wizard's chamber to create a veil of secrecy and mystery to his persona, emerging digital technology is raising new legal questions regarding the misrepresentation of identity in society.

The emergence of new technology destabilizes the legal/ethical status quo. In *The Future of Reputation*, Solove (2007) states that technology hasn't created new questions; rather it has simply made these questions more complex, bringing to light the need to adapt current law and regulations to accommodate changes brought about by technology. Legal experts recognize that information technology has changed the way we regulate and approach the application of law and established precedent. Fertik and Thompson (2010) argue that the "Internet today resembles the Old West of American history" (p. 3). While it presents numerous opportunities to strike it rich, there are also dangers in this "digital frontier."

This chapter examines differing viewpoints on how social media creates a more transparent society; whether voluntarily or involuntarily, and how this increased transparency has both positive and negative effects for individuals, companies, and governments.

The emergence of social media technology has raised new ethical concerns regarding transparency and its counterpart, privacy. Social media and Web 2.0 technologies—platforms that enable participatory sharing of content through the Internet—have altered something most people value deeply, an expectation of privacy, and made information more open and freely available. Through this open sharing of information, social media websites provide a greater level of transparency of companies and government behaviors—both good and bad.

As other chapters in this book focus on topics surrounding the idea of transparency, the primary focus of this chapter is the examination of how different social media technologies enable the misrepresentation of identity and how legal doctrine has begun to adapt to the introduction of these technologies. In doing so, this chapter presents relevant case studies and actions taken to protect online reputation and to effectively regulate misrepresentation.

## A Transparent Society

In *The Transparent Society*, Brin (1998) predicts how new technologies would force a choice between greater access to information and personal privacy:

"Light is going to shine into nearly every corner of our lives" (p. 9). Brin's primary focus was on surveillance technologies and his call for a "reciprocal transparency" to prevent government from becoming "Big Brother." YouTube was introduced in 2005, allowing users to upload and share their own videos, as well as view other user-submitted videos (Lidsky, 2010). To date, most legal issues revolving around YouTube involve such violations as defamation, privacy, or copyright law and the Digital Millennium Copyright Act, which "strengthened the legal protection of intellectual property rights in the wake of emerging new information communication technologies, i.e., the Internet" (Indiana University, 2009).Tapscott and Williams (2011) recently focused on social media's effect on privacy laws and regulations: "Privacy and data protection laws ensure that organizations collect, use, retain and disclose our personal information in a confidential manner. But today the biggest threat is not other organizations, it's us."

Tapscott and Williams's statement carries great validity with the emergence of the Web 2.0 era. With the potential loss of privacy, however, have come greater gains in transparency. YouTube and other video-sharing websites undermined the fear of government as "Big Brother" and shifted the power of video surveillance to the average citizen. Powerful technology in the hands of common individuals has given them the power to document events and publish video through social networks, adding a new level of transparency to society, but also affecting social conceptions of personal privacy. The rapid pace with which these new digital applications have emerged and been adapted has far outpaced societal ability to establish an ethical consensus to guide us in questions of appropriate disclosure.

In *The Right to Know: Transparency for an Open World*, Columbia Business School professor, Joseph E. Stiglitz (2007, p. 5), defines transparency as "the degree to which information is available to outsiders that enables them to have informed voice in decisions and/or to assess the decisions made by insiders." As we continue to examine transparency on the Internet, the word *decision* plays a vital role into how transparency occurs online.

## Voluntary Versus Involuntary

Social media tools like blogs and sharing websites have created a greater level of transparency and paradigm of openness. Social media and transparency, particular company blogs discussing their brand, or a product or service they offer, essentially mean that consumers trust that the content being posted is honest, unbiased, and truthful, when referring to that particular topic or subject. By creating and contributing to an online community, the company has elected to participate and provide information about its brand, product, or service. This is a *voluntary* approach to providing information or data. However, this does not mean social media is a truth serum and that companies or governments engage in full disclosure. There are different levels of transparency and it is at the discretion of the group or company as to how much it wants to disclose. During his campaign for president, Barack Obama promised to create a more transparent government while in office. Nevertheless, there are times when issues must be discussed behind closed doors and kept private from the public's eyes and/or ears, but this information manages to be leaked out to the public.

The act of *involuntary* transparency can be defined as information becoming public not by one's choice. Blogging has brought a greater transparency to society and has forced individuals, companies, and the government to react immediately to situations. An ethical debate in journalism revolves around the validity of information provided on the Internet, which forces journalists to be more cautious with what information they pull from the Internet. A well-known case in point involved CBS anchor Dan Rather and the Killian documents controversy, later dubbed "Rathergate."

Controversy arose during the 2004 presidential election concerning President George W. Bush's military service during the Vietnam War. During a September 8, 2004, *60 Minutes* broadcast, Rather presented documents allegedly taken from the files of the late Lieutenant Colonel Jerry B. Killian, Bush's commander, criticizing Bush's service. "President Bush had evaded the draft, used influence to join the Texas Air National Guard, and later used that same influence to whitewash his record..." (Eberhart, 2005). A forum posting at FreeRepublic.com suggested the documents were forgeries, questioning the authenticity of the documents presented during the television newscast, including the type style used on the document. Individual bloggers began their own investigation by seeking out experts of typewriter styles from the 1970s in an attempt to discover if the documents included in the CBS report were authentic or if they had actually been typed on a computer. It was later determined that the documents had been forged by their source, leading to Rather's expedited retirement from CBS (Associated Press, 2004).

Anyone is able to publish through the Internet, so transparency is no longer created at the discretion of companies and government. The power of social media allows individuals to publish what they want, about whomever they want, whether they want that information to go public or not. Blogs have become a popular publishing tool for individuals, because they are easy to setup and provide an outlet for anyone who wants to have his or her voice heard. Over the years, society has relied on blogs to provide information on niche subjects with the intent to add a unique perspective traditional media outlets may not provide or allow the community of followers to seek out the truth on particular topics, such as "Rathergate." So what happens when someone discloses information or private documents not intended to go public? The WikiLeaks release of confidential U.S. foreign policy documents is a perfect example of involuntary transparency. "Whether we like it or not, the new, involuntary transparency calls for a new code of behavior, on [sic] dictated by the reality that we can never assume we are alone or unwatched"(Bennis, Goleman, & O'Toole, 2008, pp. 17–18).Transparency advocates, like WikiLeaks' Assange, seek to create a more open and transparent society in an effort to drive greater ethical behavior among companies and government in the belief that their efforts will bring more good than harm. Similar occurrences will force governments and corporations to reassess their crisis communication plans, adopting policies and strategies for dealing with leaks of company information and how they will respond to it in a timely manner.

## Government 2.0

Shortly after President Obama took office, he issued two orders directed toward bringing greater levels of openness and transparency to the government. The first *Memorandum for the Heads of Executive Departments and Agencies* addressed the Freedom of Information Act (FIOA) and instructed that the "government should not keep information confidential merely because public officials might be embarrassed by disclosure, because errors and failures might be revealed, or because of speculative or abstract fears" (Obama, 2009a). The president's second *Memorandum for the Heads of Executive Departments and Agencies. Subject: Transparency and Open Government*, stated the following:

> My Administration is committed to creating an unprecedented level of openness in Government. We will work together to ensure the public trust and establish a system of transparency, public participation, and collaboration. Openness will strengthen our democracy and promote efficiency and effectiveness in Government. (Obama, 2009b)

The president planned to accomplish this by harnessing "new technologies to put information about their operations and decisions online and readily available to the public" (Obama, 2009a). The new technologies he was eluding to are social media tools. President Obama's adoption of social media tools was evident during his presidential campaign in 2008, where he tapped the social networks for fundraising and gathering support, which helped to defeat Democratic primary candidate Hillary Clinton and then Republican party nominee John McCain. On election night, the Obama campaign emailed its huge database of supporters, thanking them for their support and informing them how "we have a lot of work to do to get our country back on track, and I'll be in touch soon about what comes next" (Carr, 2008). The use of social networking tools gives government officials direct access to their supporters, as well as supporters having better access to government officials.

The previous example's level of social media usage may offer greater opportunities for transparency, but it does not always equate to greater transparency. According to the results of *The Foresee/Nextgov Government Transparency Study*, the American public does not believe the government is being completely open and transparent. The survey randomly questioned 5,107 U.S. citizens between August and September of 2010, where they were "asked to grade one of the following federal government entities: the White House, Congress, federal agencies and departments overall, or the federal government overall" (Freed, 2010). Based on a 100-point scale, the White House received a score of 46. While this does represent a poor score, it was not the lowest score, with government overall scoring 42, federal agencies and departments scoring 40, and Congress scoring 37 (Freed, 2010). Furthermore, out of the survey respondents, 11% gave an excellent transparency rating (scores of 80 or higher), while 67% gave a poor transparency rating (59 and lower). As the report later states, "the executive branch is the only branch of government making any proactive or quantifiable attempts at transparency" (Freed, 2010), but there is always room for improvement with scores this low for any of the government entities.

## Misrepresentation and Impersonation

"The freedom offered by the Internet truthfully to disclose information about oneself paradoxically also allows us to engage in creative self-presentation, misrepresentation and outright lies" (Whitty & Joinson, 2009, p. 55).Since the advent of the Internet and public access, individuals have been allowed to self-identify through "pseudonames" or "avatars" that do not really represent their actual person. Popular instant messaging services allowed users to create pseudonames to go by in order to have synchronous discussion online. Individuals had the discretion to determine how much information they wanted to disclose, real name included, and who they disclosed that information to. In 2003, Linden Lab introduced Second Life, a virtual world where individuals can create avatars and interact with other registered avatars. Kansas State University professors Hope Botterbusch, a specialist in legal and ethical issues in virtual environments, and Dr. Rosemary Talab, coordinator of the Kansas State University Intellectual Property Informational Center, authored "Ethical Issues in Second Life"(2009), where they bring to light the many ethical and legal questions that have arisen with Second Life, including "multiple identities" and "identity deception." Ironically, in the introductory video on their website, they promote Second Life as an opportunity to "Be Different" and "Change your Look."

Identity plays a vital role in the virtual environment when it comes to interacting and communicating with others. While the aforementioned examples are demonstrations of living through fictitious identities, there are instances where individuals try to misrepresent themselves and impersonate others online. This first became apparent with online dating websites. A study of online dating sites by Hall, Park, Song, and Cody (2010) examines how users can create their own self-description for online dating sites. Self-presentation through deception was considered to be "the conscious and intentional misrepresentation of characteristics about oneself" (p. 118). The advent of social network websites like Myspace and Facebook, as well as microblogging websites like Twitter, has caused individuals to be able to not only begin to establish their digital identity but also find new forms of misrepresentation and impersonation. One end of the spectrum of deceit and misrepresentation in social media delves into accounts of high-profile individuals where society is not aware that the accounts are run by people working for those individuals, also known as "ghost writing." On the other end, we hear about accounts being created to impersonate fictional or high-profile individuals without their permission. The proceeding sections examine the unethical use of popular social media websites Facebook and Twitter for impersonation.

## "Catfish": A Documentary of Cyber Romance and Online Anonymity

The documentary "Catfish"(Schulman & Joost, 2011) premiered at the 2010 Sundance Film Festival, focusing on events in the life of Nev Schulman, a 24-

year-old photographer, and his newfound friendship with an 8-year-old Michigan girl named Abby, her family, and a cyber romance through Facebook with Abby's older sister. (Note: Spoiler ahead.)

Nev's interaction begins when he receives a painting in the mail of one of his photographs that appeared in the newspaper and a letter from Abby telling him about how she painted his photo for him. Nevis very impressed with the quality of the artwork coming from an 8-year-old; he begins to build a pen pal relationship with Abby, sending her letters and more photos to paint. After a few correspondences, Abby's mother, Angela, contacts Nev to thank him for supporting Abby's artwork and sending more photos for Abby to paint.

The relationship grows and Nev decides to become Facebook friends with Angela, other members of her family, and some family friends. What begins as a simple correspondence later escalates into a cyber romance when Nev "friends" Megan, Abby's 19-year-old sister, and their long-distance relationship begins to flourish through e-mail, online chats, and phone calls. Ariel Schulman, Nev's brother, and friend Henry Joost, were the individuals filming Nev's life and both saw a story developing with this cyber romance. Later, suspicions arose when stories Megan was telling didn't seem to match up, from songs ripped off the Internet that she claimed to write to search results coming up empty on Abby's alleged gallery shows. On a business trip to Colorado, the three men decide to get to the bottom of the story and take a detour to Michigan to meet Megan and her family. It is on this trip where they discover the address to the farm Megan told Nev she bought to raise horses was not owned by Megan, and all of the letters Nev sent to that address were in the mailbox with "return to sender" stamps on them.

The next day they travel to Abby and Angela's residence to meet the family, only to discover neither of them look like their Facebook profile photos. After spending a few hours together, Nev confronts Angela, where he finds out Angela fabricated a majority of her stories and her accounts on Facebook. She confesses to creating the 15 semi-fake accounts, each account representing a fragment of her own personality, which were created to help her avoid the stresses of her everyday life. All controlled by Angela, the accounts use some family members' names, but the photos were taken from other accounts, so they had attractive representations. Nev also learns Angela does have a daughter named Megan, but she hasn't lived in the house for a long time. The woman whose photos were used to represent Megan, Aimee Gonzales, is a photographer/model from Seattle, Washington, who is married with two children. Angela admits to creating the characters in pursuit of friendship.

Although not every detail of the movie is documented in this section, this brief overview illustrates the point of deception and misrepresentation through the use of social media. Before proceeding, it is worth noting the controversy surrounding this documentary and whether it can be considered a true documentary or whether it is actually considered a feature film (Gardner, 2010; Ehrlich, 2010; Zeitchik, 2010). Did Nev and his colleagues suspect or perhaps realize all along that this was a fake account but continued filming because it would make a good movie? (Gardner, 2010) Scripted or not, the film still provides an example of how social media tools can be used to create accounts that do not truly represent the individual who runs the account. According to the *Facebook Statement of Rights and Responsibilities*, what Angela was portrayed doing violates numerous guidelines found under "Registration and Account Security." This section states, "Facebook users provide their real names and information…" and later goes on to say, "You will not provide any false personal information on Facebook, or create an account for anyone other than yourself without permission" ("Statement of rights," 2010).

While this documentary highlights a cyber romance similar to classic doctored online dating profiles, it paints a larger picture of the effects of social media and deception, as well as demonstrates how one person can manipulate a fictional reality through multiple personas.

## Celebrities, Companies, and Trademark Infringement

While individuals can be criticized for unethical misrepresentation through social networks, other forms provide opportunities for others to spoof or parody celebrities and commercial brands through impersonation. There have been a number of cases brought to the courts related to questions surrounding parody accounts on Twitter and allegations of trademark infringement.

In June 2009, St. Louis Cardinal manager Tony La Russa filed a complaint in the Superior Court of California against TWITTER, Inc., alleging that the defendant maintains a site allegedly owned by La Russa, containing his photograph and written statements implied to have been made by the plaintiff. The statements are claimed to be highly derogatory, suggesting alcohol abuse, and damaging to La Russa's reputation and his trademark rights. Moreover, the complaint contended that the domain name of the site, and the use of the plaintiff's name, dilutes the plaintiff's trademark, by creating the misleading impression that La Russa authors the site, and that the site encourages individuals to become members in order to receive Twitter messages.

The case was removed from the Superior Court's calendar in October 2009, with the news media reporting various conflicting terms of settlement. Lacking an official disposition of the case, PCWorld offered the following insights into the weaknesses in La Russa's case. First, La Russa should know that celebrity is going to generate parody through imitation. This is not very different from *Saturday Night Live*, although it reaches a much narrower audience. Still, Twitter has an anti-imitation process for reporting and removing attempts at impersonation, which La Russa did not pursue. Second, that same policy, however, states that parody impersonations are permissible if clearly labeled as parody. Third, online service providers are shielded from responsibility for third-party expression. Section 230 of the Communications Decency Act of 1996 prohibits a claim such as La Russa's (Raphael, 2009).

While LaRussa sought legal remedy through libel and privacy law, corporations have also attempted to take legal action against social media websites through trademark law. ONEOK Inc., a natural gas company in Oklahoma, filed suit against Twitter on counts of trademark infringement and contributory infringement, alleging that an anonymous user had created and employed an imposter Twitter account using the name ONEOK and the corporate trademark. Its complaint alleges that the imposter account had tweeted corporate information, which the company held to be potentially damaging to ONEOK's public image in the investor community and energy industry and could negatively impact the company's shareholders. ONEOK sought to obtain the imposter's identity and contact information and to have the ONEOK account assigned to the company. Twitter refused to do so in both instances, at which point ONEOK filed suit, seeking a permanent injunction against further use of the account (ONEOK, 2009).

The case was dropped by ONEOK, perhaps following the removal of the site by Twitter. Issues such as these may be the motivation behind Twitter's decision to introduce "Verified Accounts," through which celebrities, athletes, public officials, and so forth, can display "verified account" status on their Twitter pages. This allows those notables to distinguish their legitimate sites from impersonators. Twitter is said to be considering a similar approach for businesses (ONEOK, 2009).

The previous cases involve instances of parody, representing a form of social commentary and, as such, demonstrating a need to protect the user's identity as being necessary for free speech. These cases can be seen as being less manipulative than the type of situation portrayed in "Catfish," and therefore, at least ethically equivocal. Public figures can take care of themselves. The need to protect the identity of the user takes on an even higher priority when we examine the role of social media in providing a forum to speak out against the politically powerful.

## Social Media's Role in Politics

Geoff Livingston, cofounder of the communication consultancy *Zoetica*, discusses how social media has become the new battleground for politics. "With new media at hand, elections become a time for innovation, and online engagement can lead to enormous influence" (Livingston, 2010). Politicians and their followers have taken advantage of the power of social media in unethical ways to influence others around election time. The following case studies take a look at how social media has played a role in politics and different legal issues.

Just like the accounts of users creating their own self-description for online dating websites to persuade other users of their persona, similar incidents have occurred in the political arena to gain an edge over competitors.

During the 2010 mayoral race in Toronto, Canada, the local media covered a story alleging that the staff of Rob Ford, Toronto's recently anointed mayor elect, was using a fake Twitter account (@QueensQuayKaren) to convince an individual who had incriminating video evidence of Ford into handing it over to Ford's staffers (Fleet, 2010). Fraser Macdonald, a Ford staffer, created the account to impersonate a supporter of the rival candidate George Smitherman, and he befriended the individual in possession of the audio recording of Ford looking to purchase prescription painkillers on the street. After receiving a copy of the tape, campaign members leaked it to the *Toronto Star*, a Ford-friendly newspaper, rather than allowing it to get into the hands of other media outlets, where it most likely would have hindered Ford's chances of election. Although the account was deleted once mainstream media caught wind, numerous outlets captured screenshots of the "tweets" that were sent out. Macdonald was also documented bragging about the success of the deception on his personal Twitter account (@macdonaldfraser), which read as follows: "Enjoying the feedback, positive and negative, re: one of the many ways the Ford campaign outsmarted the competition. #voteto#getoverit."

While the incident in Toronto illustrates how a Twitter account can be setup internally to gain a competitive edge for an elected official, there are times accounts are created to represent a politician without the individual's knowledge. According to Nikita R. Stewart (2011) of the *Washington Post*, a fake account appeared for Washington, D.C., Mayor Vincent C. Gray and questions arose in regard to who was running the account. In a separate case, Morrissey (2011) of the *DCist* notes that this is a perfect example of how people can easily be deceived: "A Councilmember who actually worked alongside the new Mayor for the last four years didn't realize the fake account wasn't him, there's certain to be others out there."

Despite the fact that no legal action was pursued in the previous two cases, individuals have filed lawsuits in other online deception cases. The goal of these lawsuits was to gain control of their online identity and protect their reputation. However, questions arise when it comes to enforcing law in an environment without physical boundaries.

Conversely, politicians have been aggressive in combating the use of social media to parody their actions and policies. In May 2008, Larry Dominick, town president of Cicero, Illinois, filed a Petition of Discovery in Cook County Circuit Court, seeking the identity of the Myspace user who had created two false profiles of him. His petition requested a court order requiring Myspace to reveal the name of the user, claiming "defamation, invasion of privacy, and related torts arising from various statements on a social networking website known as Myspace" (Amicus Brief, 2008, p. 1). The petition continued: "These imposter profiles falsely state that they are the profiles of Petitioner, falsely state that they were created, posted, and published by Petitioner, and include defamatory matter concerning the Petitioner" (Amicus Brief, 2008, p. 1). Although Dominick provided no details to support his claim that the statements about him were false, Myspace removed the pages in question.

On June 4, 2008, the Electronic Freedom Foundation (EFF) responded, arguing in an amicus brief that the First Amendment protects anonymous speech unless Dominick could demonstrate a viable defamation claim. News stories characterized the content as "silly juvenile stuff" and indicated that retribution was a motive in seeking the poster's identity (Noel, 2008). Furthermore, the brief argued, the Federal Stored Communications Act prohibits government officials from seeking consumer identities through the civil discovery process. The EFF noted that a decision by this court must consider the precedential implications on the rights of Internet users throughout the state and continued on to express concern that discovery procedures might be used by government officials to "harass, intimidate, or silence critics in the public forums opportunities presented by the Internet" (Amicus

Brief, 2008, pp. 2–3). Mr. Dominick's petition failed to meet the strict Constitutional guidelines of established case law.

Dominick withdrew his petition on June 13 without stating his reasons for doing so. The EFF stated publicly that Dominick failed to provide specific allegations of defamation or any proof that the allegedly defamatory statements were false.

In a similar case in May 2010, Pennsylvania's Attorney General Tom Corbett requested an investigative grand jury to issue a subpoena requesting that Twitter respond by providing the identities of two anonymous Twitter account users who had criticized Corbett as he competed in the Republican primary for that party's gubernatorial nomination (Subpoena of Twitter, 2010). A spokesman for Corbett indicated that the subpoena stemmed from a criminal case involving a former political aide convicted in a political scandal. The subpoena was issued on May 6, 2010.

Free speech advocates criticized Corbett as an abuse of his official powers as attorney general to silence critics through political retaliation. "Anonymous speech is a longstanding American right," (Schwartz, 2010) said Paul Alan Levy of the Public Citizen Litigation Group, a Washington organization that has defended anonymous commenters who have been sued by companies and politicians. Witold Walczak, legal director for the American Civil Liberties Union of Pennsylvania, said in a statement that "any subpoena seeking to unmask the identity of anonymous critics raises the specter of political retaliation" (Schwartz, 2010).

One of the targets of the subpoena responded to the request in his blog: "That's not a question I'm prepared to answer. We intend to remain anonymous and we will not confirm or deny who we are, whether we're connected to the defendants. What we have to say to Tom Corbett, we have said in our blog and we will continue to say it on our blog, because what happens to us could happen to anyone. This is really not about the identities of these particular bloggers. This is about the right of an American to anonymously criticize a public official" (Tom Corbett, 2010). Corbett withdrew the subpoena on May 19, 2010. He offered no comment as to why he withdrew the order, responding only that the subpoena had been part of a grand jury investigation and that he could not offer comment beyond that.

The various cases examined here demonstrate that impersonation, misrepresentation, and anonymity can be seen as highly unethical or an ethical necessity, depending on the circumstance. Impersonation through social networks can be used to manipulate others, emotionally or perhaps financially, and is therefore, highly unethical. Impersonation can also serve as a form of parody, offering a satire of famous faces or familiar products or services. Satire has long been seen as an acceptable form of social commentary and part and parcel of an open society. Only the technology is new in the examples offered. Social media also offers a powerful equalizing force, allowing individuals to express and publish their opinions, communicate with like-minded individuals, and seek to influence public debates. In these instances, anonymity allowed people the social space to speak truth to power. As such, the ability to remain anonymous is an essential right, but one that may be threatened by proposed legislative action seeking to regulate social media.

## Movements for Regulating Cyberspace

In 2006, David Johnson and David Post published the article "Law and Borders: The Rise of Law in Cyberspace" on the *First Monday* website. They discussed how global computer-based technologies now ignore traditional territorial boundaries and new boundaries separated by usernames and passwords create the boundaries of cyberspace. These new boundaries now threaten traditional laws based on the idea of territorial boundaries and the redefinition of the role of law. The online environment and culture have changed so fast, they haven't allowed the law to catch up. Referring back to Fertik and Thompson's (2010) reference of the digital frontier and the Wild West, "there is no sheriff in town, and Internet users have been left with rough frontier justice" (p. 5)

To date, courts consider online impersonation as a form of identity theft; however, the Federal Trade Commission (FTC) issued the Fair and Accurate Credit Transactions Act (FACTA), which defines identity theft as "a fraud that is committed or attempted, using a person's identifying information without authority," and later goes on to define identifying information as "means of identification" in the federal criminal statute. ("Fair and accurate," 2003). Since most cases of online impersonation would not qualify as a "means of identification," and usually there is no money involved, district attorneys would not prosecute the perpetrators. This gives free reign for individuals wanting to impersonate others, without any true repercussions.

In recent years, there have been attempts to begin pushing for greater regulation when it comes to online impersonation. In September of 2009, the state of Texas enacted H.B. No. 2003, making it a third-degree felony to impersonate individuals on social networking websites "relating to the creation of the offense of online harassment" (H. B. No. 2003, 2009). In addition, this law also applies to text messages. Specifically, the opening to Section 33.07 states the following:

> A person commits an offense if the person uses the name or persona of another person to create a webpage on or to post one or more messages on a commercial social networking site: (1) without obtaining the other person's consent; and (2) with the intent to harm, defraud, intimidate, or threaten any person." (H. B. No. 2003, 2009)

As of January 1, 2011, the state of California passed State Bill 1411, making it a misdemeanor to impersonate another individual online.

> This bill would provide that any person who knowingly and without consent credibly impersonates another actual person through or on an Internet Web site or by other electronic means, as specified, for purposes of harming, intimidating, threatening, or defrauding another person is guilty of a misdemeanor. The bill would, in addition to the specified criminal penalties, authorize a person who suffers damage or loss to bring a civil action against any person who violates that provision, as specified. (S.B. 1411, 2010)

Although this maybe a step in the right direction, these bills raise potential constitutional questions in regard to anonymous speech, parody, and satire, as well as commercial speech and trademark dilution. Also, what happens when the individual responsible for the impersonation is from a state outside of Texas or California? Larry Lessig (1996) argues, however, that traditional legal approaches are shortsighted and ineffective.

## Code as Law

In a recent American Bar Association convention keynote address, Harvard Law Professor Larry Lessig summed up his frustration toward legislators and media regulators who still do not grasp the degree to which the Internet has "fundamentally changed the exchange of shared content." In a speech titled "Code Is Law, Does Anyone Get It Yet?," he lambasted their approach as one that "forces traditional regulatory models without understanding or regard for the impact of potential market, architectural and social norms." Such an approach, according to Lessig, is "ineffective, inefficient and naïve" (cited in Zahorsky, 2011):

> Our first instinct is to invoke the law quite forcefully....The right instinct would be to modify the law and the market to reflect the new ways innovation and technology are being used, while also making sure artists are getting paid....The law needs to dereg-

ulate a certain area of culture in order to effectively regulate where it should properly be applied.

Lessig (1996, p. 1403) has been addressing the futility of the application of traditional legal constraints in cyberspace since 1996: "The architecture of cyberspace compels a different kind of regulation." Responding to earlier arguments of Johnson and Post (1996): "I think Johnson and Post are most ambitious, one might say romantic," in constructing a vision of "a democracy in cyberspace—a world of cybercitizens deciding on the laws that apply to them" (Lessig, 1996, pp. 1403–1411). At that time, Lessig noted, cyberspace was a place of relative freedom, fairly crude technologies of constraint, and what controls did exist were exercised through the creation and enforcement of social norms. Regulation in cyberspace, Lessig argued, would be different. If one seeks to induce a certain behavior, they do not need to induce conformity through coercion. One only needs to change the code—the software that defines the terms by which people gain access to the system or use assets provided by the system. In short, Lessig argues that law would be inscribed increasingly in code. Such a code is an "efficient means of regulation," (p. 1403) in that it is obeyed because one can do nothing else. He contends that "Law as Code" was the beginning of a "perfect technology of justice, one that allows policy makers to select a social end, and then assure compliance by individuals to that end. Code as software becomes a means to that end" (p. 1410).

Such an approach can be found in the attempt by Twitter to "verify" the authenticity of a message source by identifying and confirming particular accounts as Verified Twitter accounts. This would signify that "celebrities, musicians, athletes, actors, public officials and public agencies on the service can now display a 'verified account' button on their Twitter pages" (Cashmore, 2009). "The goal of this program is to limit user confusion by making it easier to identify authentic accounts on Twitter" (Twitter, n.d.). These types of movements, bringing visibility to authentic accounts, will assist in providing the public a way to verify accounts and avoid instances of misrepresentation and confusion without enacting a law.

Twitter suspended this effort in September 2010 as it came to discover that identity confirmation is more challenging than previously thought. Who would determine who owns what brand or name? Can a small business get verified? What about very common names? Indications are that Twitter is working on developing a new approach to verification that seeks to expand the benefits beyond the categories of users identified earlier (Cattachio, 2010).

# Conclusion

"Transparency is not about eliminating privacy. It is about giving us the power to hold accountable those who would *violate* it" (Brin, 1998, p. 334). The Internet and social media tools have created a double-edged sword in regard to privacy and staying connected with today's society. While we must remind ourselves that these tools have changed the way we communicate, how each of us views privacy and transparency in today's society also varies. With the evolution of new technologies on a daily basis, it is challenging to regulate the new "digital frontier" (Fertik & Thompson, 2010). The challenge in applying traditional law to cyberspace is that the online environment disregards any geographical boundaries and the actions performed online are almost entirely independent of a set physical location. While this has made it easier as a society to stay connected, the notion of applying clear legal rulings becomes harder to govern with traditional laws due to this boundary-less environment. How effective are the laws passed to help regulate social networking websites and impersonation online in the states of Texas and California has yet to be seen. While movements like "verified accounts" on Twitter did not establish absolute law on the Internet, they begin to provide a higher level of protection to individuals' personas, similar to approaches suggested by Lessig. The upcoming years will be interesting as we continue down the yellow brick road in pursuit of The Great Oz to provide us with answers in regulating the issues surrounding transparency and misrepresentation through social media websites.

## References

Amicus Brief, *Larry Dominick v. Myspace, Inc.*, Electronic Freedom Foundation, Circuit Court of Cook County, Illinois County Department Law Division, June 4, 2008.

Associated Press. (2004, September 21). *CBS news admits Bush documents can't be verified.* Retrieved from http://www.msnbc.msn.com/id/6055248/ns/politics/

Bennis, W., Goleman, D., & O'Toole, J. (2008).*Transparency: How leaders create a culture of candor.* San Francisco, CA: Jossey-Bass.

Brin, D. (1998). *The transparent society: Will technology force us to choose between privacy and freedom?* Reading, MA: Addison-Wesley.

Carr, D. (2008, November 9). *How Obama tapped into social networks' power.* Retrieved from http://www.nytimes.com/2008/11/10/business/media/10carr.html

Cashmore, P. (2009, June 11). *Twitter launches verified accounts.* Retrieved from http://mashable.com/2009/06/11/twitter-verified-accounts-2/

Cattachio, C. (2010, September 28). *Twitter officially halts verifying accounts, working on new system.* Retrieved from http://thenextweb.com/socialmedia/2010/09/28/twitter-officially-halts-verifying-accounts-working-on-new-system/

Eberhart, D. (2005, January 31). *How the blogs torpedoed Dan Rather.* Retrieved from http://archive.newsmax.com/archives/articles/2005/1/28/172943.shtml

Ehrlich, D. (2010, December 4). *"Catfish" lawsuit may expose truth behind documentary.* Retrieved from http://blog.moviefone.com/2010/12/04/catfish-lawsuit/

Fertik, M., & Thompson, D. (2010). *Wild west 2.0: How to protect and restore your online reputation on the untamed social frontier.* New York: American Management Association.

Fleet, D. (2010, October 30). *Unethical social media at its worst: Rob Ford's fake Twitter account* [Weblog message]. Retrieved from http://socialmediatoday.com/davefleet/227540/unethical-social-media-its-worst-rob-ford-s-fake-twitter-account?utm_source=feedburner&utm_medium=twitter&utm_campaign=Social+Media+Today+%28all+posts%29&utm_content=Twitter

Freed, L. (2010, October 21). *ForeSeeresults/Nextgovgovernment transparency study.* Retrieved from ForeSee Results website: http://www.foreseeresults.com/research-white-papers/_downloads/foresee-results-nextgov-government-transparency-study-q3-2010.pdf

Gardner, E. (2010, December 3). *Exclusive: New lawsuit seeks to expose truth behind "Catfish"* [Weblog message]. Retrieved from http://www.hollywoodreporter.com/blogs/thr-esq/lawsuit-seeks-expose-truth-catfish-55969

Hall, J. A., Park, N., Song, H., & Cody, M. J. (2010). Strategic misrepresentation in online dating: The effects of gender, self-monitoring, and personality traits. *Journal of Social and Personal Relationships, 27*(1), 117–135. doi:10.1177/0265407509349633

Harmon, J. (2009, September 9). *Symbolism of the "Wizard of Oz"* [Weblog message]. Retrieved from http://wickedtour.net/symbolism-of-the-wizard-of-oz

H.B. No. 2003. (2009, May 23). *Act relating to the creation of the offense of online harassment.* Legislature of the State of Texas. Retrieved from http://www.legis.state.tx.us/tlodocs/81R/billtext/html/HB02003F.htm

Indiana University. (2009, July 20). *What is the Digital Millennium Copyright Act?* Retrieved from http://kb.iu.edu/data/alik.html

Johnson, D., & Post, D. (1996). Law and borders: The rise of law in cyberspace. *Stanford Law Review, 48,* 1–20.

Lessig, L. (1996). The Zones of Cyberspace. *Stanford Law Review, 48,* 1403–1411.

Lidsky, D. (2010, February 1). *The brief but impactful history of YouTube.* Retrieved from http://www.fastcompany.com/magazine/142/it-had-to-be-you.html

Livingston, G. (2010, September 23). *Social media: The new battleground for politics.* Retrieved from http://mashable.com/2010/09/23/congress-battle-social-media/

Morrissey, A. (2011, January 28). *The Twitter gap.* Retrieved from http://dcist.com/2011/01/the_twitter_gap.php

Noel, J. (2008, May 17). *Cicero town president wants Myspace poser's identity revealed.* Retrieved from http:// http://articles.chicagotribune.com/2008-05-18/news/0805180119_1_myspace-social-networking-sites-town-president

Obama, Barack, to Heads of Executive Departments and Agencies regarding Freedom of Information Act, January 21, 2009a, Office of the Press Secretary, The White House. Retrieved from http://www.whitehouse.gov/the_press_office/FreedomofInformationAct

Obama, Barack, to Heads of Executive Departments and Agencies regarding Transparency and Open Government, January 21, 2009b, Office of the Press Secretary, The White House. Retrieved from http://www.whitehouse.gov/the_press_office/TransparencyandOpenGovernment

ONEOK, Inc. v. Twitter, Inc. Complaint filed in the U.S. District Court of Northern California. (2009, September 15). Citizen Media Law Project. Retrieved from http://www.citmedialaw. org/threats/oneok-inc-v-twitter

Rapael, JR (2009). Three strikes against Tony La Russa's Twitter lawsuit. PCWorld. June 4, 2009. Retrieved from http://pcworld.com/article/166151/Tony_La_Russa_Twitter_suit.html

S.B. 1411. (2010, March 25). Bill analysis. Senate Judiciary Committee.Senator Ellen M. Corbett, Chair. California State Senate. Retrieved from http://info.sen.ca.gov/pub/09—10/bill/sen/sb _1401–1450/sb_1411_cfa_20100419_123604_sen_comm.html

Schulman, A. (Filmmaker), & Joost, H. (Filmmaker) (2011). "Catfish" [DVD].

Schwartz, J. (2010, May 20). Twitter fighting Pennsylvania subpoena seeking names of 2 tweeters. Retrieved from http://www.nytimes.com/2010/05/21/technology/21twitter.html?_r=1

Solove, D.J. (2007).The future of reputation. New Haven, CT: Yale University Press.

Statement of rights and responsibilities. (2010, October 4). Retrieved from https://www.facebook. com/terms.php?ref=pf

Stewart, N. R. (2011, January 27). Fake Twitter account pops up for Mayor Gray. The Washington Post. Retrieved from http://voices.washingtonpost.com/dc/2011/01/fake_twitter_account_ pops_up_f.html

Stiglitz, J. E. (2007). Foreword. In A. Florini (Ed.), The right to know: Transparency for an open world (p. 5). New York: Columbia University Press.

Subpoena of Twitter, Inc. Commonwealth of Pennsylvania Statewide Investigating Grand Jury. Supreme Court of Pennsylvania. May 6. 2010.

Tapscott, D., & Williams, A.D. (2011, January 1). Social media's unexpected threat [Weblog mes- sage]. Retrieved from http://www.theglobeandmail.com/report-on-business/commentary/ don-tapscott/social-medias-unexpected-threat/article1854656/

Twitter. (n.d.). About verified accounts. Retrieved from http://support.twitter.com/groups/31- twitter-basics/topics/111-features/articles/119135-about-verified-accounts

United States Congress, (2003). Fair and accurate credit transactions act of 2003 (PUBLIC LAW 108–159). Retrieved from website: http://www.gpo.gov/fdsys/pkg/PLAW-108publ159/pdf/ PLAW-108publ159.pdf

Whitty, M. T. & Joinson, A. N. (2009). Truth, Lies and Trust on the Internet. Psychology Press. New York, NY.

Zahorsky, R. (2011, April 12). Regulators ignoring internet's real effects, Lessig says, and free exchange suffers. American Bar Association Journal: Law News Now. Retrieved fromhttp://www.abajournal.com/news/article/regulators_ignoring_internets_real_effects_less ig_says_and_free_exchange_su/

Zeitchik, S. (2010, September 19)."Catfish" blurs line between documentary and feature film. Retrieved from http://articles.latimes.com/2010/sep/19/entertainment/la-ca-catfish-20100919

# · 9 ·

# Digital Red Light Zones

## Alternative Approach to Regulating Adult Online Social Media

BRUCE E. DRUSHEL

## Introduction

The regulation of media content that is offensive and inappropriate for some audiences but not obscene has been problematic as a matter of history. The U.S. Supreme Court first affirmed that obscene speech was outside of the protection of the First Amendment in its 1957 *Roth v. United States* (1957) decision. The Court subsequently ruled that some material that wasn't obscene nevertheless could be kept from children, so long as adults retained access (*Ginsberg v. New York*, 1968). In so doing, the Court effectively broadened the legal concept of *channeling* to include media content to address social changes in the 1960s that were rendering Western culture less and less homogeneous and more and more prone to internal disagreement in its assessment of what was moral, acceptable, and tasteful.

Channeling long has been accepted as a constitutionally consistent method for regulating potentially offensive situations, such as barking dogs, panhandling, and the operation of adult book stores, by placing restrictions on their permissible time, place, and manner (TPM). Cities have used zoning codes, for example, to restrict the location of adult book stores or their concentration in a particular area; barking dogs, garbage trucks, and other creators of noise are not permitted at night when most people are trying to sleep; and panhandlers may beg for money but may not do so aggressively or at transit stops or automated teller machines. If the more traditional media represented an abstracted version of this reality, where magazines, radio personalities, and videos could be psychic nuisances, the emerging media, including online social networks, potentially represent yet another new environment for the application of this approach. Recently, the Internet Corporation for Assigned Names and Numbers (ICANN), the agency that administers the registration of internet domain names, announced it was considering adding .xxx to the list of existing domain suffixes, which also include .com, .net, and .org ("Porn sites," 2010). The proposal seems to resolve constitutional issues that have plagued previous attempts at regulating potentially offensive online content, since it would be left to internet users whether to employ software to block sites with the suffix, and since its adoption by domain name owners would be voluntary. Thus sites, including online social networking sites, adopting the .xxx suffix may have less to fear from would-be regulators; other sites would less likely be inadvertently blocked by overly sensitive software, and users may find adult sites easier to locate and themselves less intimidated in designated adult areas of cyberspace. On the other hand, its effectiveness is in doubt, since the company that requested the suffix predicted just one-tenth of the adult websites on the internet would choose the suffix (Associated Press, 2010).

This chapter examines the .xxx domain suffix proposal as a potential tool for the control of indecent speech in cyberspace and particularly in the burgeoning area of online social media. It reviews the relevant history of previous channeling attempts and, given Western culture's experience with both government-imposed and self-imposed content regimes, speculates as to the potential effectiveness of "Digital Red Light Zones" as well as unintended consequences of them.

## The Sanctity of Cyberspace

As the concept of cyberspace—a near-boundless global digital network capable of not just making traditional forms of communication more efficient but of creating new ones—began to penetrate the popular consciousness in the 1980s, much of the initial emphasis seemed to be on speed, convenience, and availability. No longer would communication have to rely upon the speed of postal services, no longer would consumers need to physically visit a retailer to make a purchase, and no longer would consumer choice be a matter of geographic location. As the convergence of technologies became the catchphrase and more portable multifunction devices with the capability of accessing a greater variety of digital communications including pictures, video, and sound proliferated, the emphasis became the ubiquity of cyberspace and its always-connected denizens. If broadcasting had become "uniquely pervasive" by the 1970s, to use the language of the U.S. Supreme Court's *Pacifica* (1978) decision, then what of social media, which may be accessed on desktop and laptop computers, as well as on digital television, smartphones, and tablet devices? These services, along with the rest of cyberspace, literally can follow their users (including, in many cases, unsupervised children from the age of 13) anywhere and typically have shown little sincerity in their efforts to self-police inappropriate content.

A survey of content restrictions for the digital social media services most widely used in the United States and Europe as reflected in their "terms of service" (TOS) pages suggests on the one hand a corporate public stance against indecent content but also a lack of specificity, making consistent enforcement difficult. Facebook and Twitter have the most specific policies, with Facebook proscribing, "nudity or pornographic material of any kind" among its advertisements and "sponsored stories," though apparently not in individual pages (Facebook, 2011). Similarly, Twitter advises users, "You may not use obscene or pornographic images in either your profile picture or user background," but apparently leaves open the question of indecent text in postings (Twitter, 2011).

At the other extreme is Google, whose current (as of March 1, 2012) TOS states the following:

> Our Services display some content that is not Google's. This content is the sole responsibility of the entity that makes it available. We may review content to determine whether it is illegal or violates our policies, and we may remove or refuse to display content that we reasonably believe violates our policies or the law. But that does not necessarily mean that we review content, so please don't assume that we do. (Google, 2012)

This publicly vague policy is in sharp contrast, at least philosophically, with an earlier version (from April 16, 2007) that warned the following:

> 8.3. Google reserves the right (but shall have no obligation) to pre-screen, review, flag, filter, modify, refuse or remove any or all Content from any Service. For some of the Services, Google may provide tools to filter out explicit sexual content. These tools include the SafeSearch preference settings....In addition, there are commercially available services and software to limit access to material that you may find objectionable.

8.4. You understand that by using the Services you may be exposed to Content that you may find offensive, indecent or objectionable and that, in this respect, you use the Services at your own risk. (Google, 2007)

LinkedIn's policy stakes out a middle ground. It instructs users they may not, "upload, post, email, InMail, transmit or otherwise make available or initiate any content that…is unlawful, libelous, abusive, obscene, discriminatory or otherwise objectionable" (LinkedIn, 2011).

Previous attempts by state actors to regulate online material that may be offensive to some and inappropriate for minors but not legally obscene generally have failed court review in the United States. The best-known attempt was The Communications Decency Act of 1996 (CDA), enacted as Title V of the larger Telecommunication Act of 1996 (110 Stat. 56). Two provisions of the CDA prohibited transmission of obscene or indecent messages to recipients under the age of 18 by phone or by interactive computer service (47 U.S.C.A. § 223(a) [Supp. 1997]; 47 U.S.C.A. § 223(d) [Supp. 1997]), subjecting violators to fines and imprisonment. Exceptions were made for those taking "good faith, reasonable, effective, and appropriate actions" to restrict access by minors, including requiring recipients to show proof of age (47 U.S.C.A. § 223(e)(5) [Supp. 1997]).

The legislation was challenged by a laundry list of plaintiffs led by the American Civil Liberties Union (ACLU). A special three-judge panel of the U.S. District Court overturned it on both First Amendment and Fifth Amendment grounds. In his contribution to a three-part opinion, District Judge Stewart Dalzell wrote that the World Wide Web (web) was a wholly new medium that constituted "the most participatory form of mass speech yet developed" and was entitled to "the highest protection from governmental intrusion"(ACLU v. Reno, 1996).

The U.S. Supreme Court agreed (though it failed to reach the Fifth Amendment issue), finding the CDA to be a content-based measure deserving of special scrutiny and distinguishing the web from broadcast radio and television (where material that is indecent but not obscene may be restricted) because it was neither invasive nor scarce. Writing for a unanimous Court, Justice John Paul Stevens found the government's concern for underage users laudable but, given the expectation that they always were among the larger community of web surfers and could not be reliably screened from offensive sites, also reaffirmed the Court's position that the possibility of a minor's being exposed to indecent material was insufficient to justify the inevitability of speakers' being chilled from engaging in otherwise lawful speech (Reno et al. v. ACLU et al., 1997).

In response to the invalidation of the CDA, Congress made two further attempts in the succeeding decade to regulate indecent speech on the web. The Child Online Protection Act (COPA), passed in 1998, imposed criminal penalties of up to 6 months in prison and a $50,000 fine on those knowingly making available for profit

> ... any communication, picture, image, graphic image file, article, recording, writing, or other matter of any kind that is obscene or that—
>
> (A) the average person, applying contemporary community standards, would find, taking the material as a whole and with respect to minors, is designed to appeal to, or is designed to pander to, the prurient interest;
>
> (B) depicts, describes, or represents, in a manner patently offensive with respect to minors, an actual or simulated sexual act or sexual contact, an actual or simulated normal or perverted sexual act, or a lewd exhibition of the genitals or post-pubescent female breast; and
>
> (C) taken as a whole, lacks serious literary, artistic, political, or scientific value for minors.(§ 231[e][6])

Like the CDA before it, the COPA did provide for an affirmative defense against prosecution for those who could demonstrate they had attempted to verify the age of underage users, including by requiring credit cards or access codes.

A challenge begun again by the ACLU eventually led the U.S. Supreme Court to affirm an appeals court decision blocking enforcement of the COPA because its restrictions on protected speech were not the least restrictive means necessary to protect underage users. The Court agreed that filtering software likely would prove more effective if only because it would work with sites outside of the United States, which were thought to account for as much as 40 percent of the target content (*Ashcroft v. ACLU*, 2004). A permanent injunction against enforcement of the COPA was issued 3 years later (*ACLU v. Gonzales*, 2007).

In the meantime, Congress passed a third measure, the Children's Internet Protection Act (CIPA; 2001) which, taking its cue from lower court decisions, required public libraries receiving grants from the Library Services and Technology Act (LSTA) or internet service discounts through the E-Rate program to use filtering software to prevent users from accessing indecent content. A coalition of groups led by the ACLU and the American Library Association sued to block its implementation, but the U.S. Supreme Court found it passed Constitutional muster. In a plurality opinion, Chief Justice William Rehnquist found filters to be an acceptable measure, given the impossibility of sorting through the volume of continuously changing information online and certainly preferable to blocking entire websites: "Because public libraries' use of Internet filtering software does not violate their patrons' First Amendment rights, CIPA does not induce libraries to violate the Constitution, and is a valid exercise of Congress' spending power" (*United States v. American Library Association*, 2003).

Further, since the Court has in the past approved of imposing some limitations on the use of government funds to facilitate programs (see *Rust v. Sullivan*, 1991), the "CIPA does not impose an unconstitutional condition on libraries that receive E-rate and LSTA subsidies by requiring them, as a condition on that receipt, to surrender their First Amendment right to provide the public with access to constitutionally protected speech." Rehnquist added that a law restricting availability of pornographic materials to library patrons is in keeping with collections practices for more traditional hard copy materials already in place at most libraries (*United States v. American Library Association*, 2003).

## Not Obscene but Not Acceptable

The sort of content Congress seemed intent on restricting has long proven problematic for regulators both definitionally and constitutionally. While the First Amendment appears to prohibit legislation restricting speech of any kind, the U.S. Supreme Court has carved out exceptions, including speech that induces panic or that poses an imminent security threat to the nation. The exception for obscene speech has formally existed since *Dunlop v. U.S.*(1897), though the *Hicklin* standard for obscenity earlier had been adapted to American cases from British common law (*Regina v. Hicklin*, 1868).

The U.S. Supreme Court first recognized "indecent" speech as a distinct category in its *Federal Communications Commission v. Pacifica* decision in 1978 and provisionally approved Federal Communication Commission (FCC) efforts to restrict it on broadcast stations to times when unsupervised children were less likely to be in the audience. The case famously addressed the broadcast of comedian George Carlin's recorded "Filthy Words" monologue by a non-commercial community radio station in New York and represented the first effort by the Commission to regulate content it considered inappropriate for audiences but which did not meet the definitive standard for obscenity recently established by the Court in *Miller v. California* (1973). *Pacifica*, too, established a standard, one that differed from *Miller* in that material that did not appeal to the "prurient interest" and that did not lack "serious literary, artistic, political, or social value" could be kept off radio and television during hours when there was a significant risk of children in the audience.

# Okay in Your Backyard—but Not in Mine

The *Pacifica* decision was distinctive both for its contribution of meaning to the disjunctive category of "obscene, indecent, or profane" and for its successful incorporation of TPM restrictions into mediated communication. The rationale behind TPM regulations on speech is that the offensiveness of some utterances is contextual: acceptable to society and necessarily lawful but only so long as they remain in their proper place. In the case of broadcast radio, the appropriate context for willful and repeated use of Carlin's "Filthy Words" was during hours when children either could be assumed to be asleep or under the supervision of their parents.

The creation of "indecency" as a form of expression distinct from obscenity appears to have occurred decades after its appearance in federal statutes. Correspondence between the FCC's first chair and a key member of Congress suggests language in the United States Criminal Code (specifically, 18 U.S.C. § 1464) forbidding obscene, indecent, or profane speech intended that the three terms be used conjunctively despite their disjunctive denotation (Rivera-Sanchez, 1994). This, of course, occurred at a time when the accepted standard for obscenity was more easily met by content addressing sexuality, even through allusion, and years before the cultural convulsions of the 1950s and 1960s created generational, class-based, and ethnic distinctions in what was considered acceptable and offensive behavior and forced the courts and the FCC to reconsider that standard. The FCC appears to have begun evolving a specific definition of indecency based upon the pervasiveness of broadcasting as early as 1964, culminating in the U.S. Supreme Court's *Pacifica* decision. Ironically, the legal meaning of the third term, "profane," never was specified by the FCC, though it conventionally is considered to refer to "swear words" and similar language generally unrelated to sexuality or bodily functions.

The Supreme Court has ruled that "time, place, and manner regulations are acceptable so long as they are designed to serve a substantial governmental interest and do not unreasonably limit alternative avenues of communication" (*City of Renton v. Playtime Theatres, Inc.*, 1986, p. 47) and developed a four-part test to evaluate their constitutionality. To pass First Amendment muster, TPM restrictions must be content neutral, meaning they may not prohibit entire classes of expression, such as drug use or sexual relations. They must be viewpoint neutral, meaning they may not apply only to those with a particular position within a class of expression. They must serve a significant government interest, meaning legislators must be able to articulate a specific rationale for controlling a type of expressive behavior; typically, the rationale involves the protection of children or other sensitive populations. Finally, the restriction must be narrowly drawn in such a way that it advances the government interest without foreclosing other contexts for the speech and leave open alternative channels of communication.

In the years following, courts were asked to consider the First Amendment implications of moves to impose similar strictures on other media and communications, including basic and premium cable channels (*Cruz v. Ferre,* 1985), cable access channels (*Denver Area Educational Telecommunications Consortium v. FCC,* 1996), adult telephone "chat" lines (*Sable v. FCC,* 1989), and adult-themed cable channels (*United States et al. v. Playboy Entertainment Group,* 2000). In each case, the Court was unconvinced the restrictions did not impede the right of adults to indecent content, though it did find remedies short of complete bans on indecency, such as requiring a credit card or other proof of majority from users of so-called dial-a-porn services, to be acceptable (*Sable v. FCC,* 1989).

In fact, The Court's approval of software filters in *United States v. American Library Association* (2003) marked a significant departure from its approach in prior offensive content cases in that the remedies, though limited to public libraries, applied to all patrons regardless of age. Since *Butler v. Michigan* (1957) in which it refused to block adult access to content determined inappropriate for minors, the Court had shied from rationales other than age as the basis for restrictions on non-obscene content. In fact, it had singled out concern for minors and for parental prerogative in the rearing of children as two justifications for upholding broadcast indecency restrictions in *Pacifica,* wherein offensive material was confined to late night hours (Docter, 1992). A third justification, protection of adults not wishing to be exposed to such content, seemed less relevant to the remedies developed by the FCC and endorsed by the Court. Similarly, cities may insist adult magazines be kept behind the counter and sold only on proof of majority. Generally, the efforts in other media failed because those media were not "uniquely pervasive" like broadcasting and thus attempts to regulate their TPM of availability went too far. In *Sable v. FCC,* for example, the Court found an outright ban on indecent phone services overly broad, particularly given evidence from FCC rulemaking that it was unlikely children could evade proof-of-age requirements (*Sable v. FCC,* 1989).

# Pervasiveness: A Reconsideration

The years since the *Pacifica* and *Sable* cases have seen the development and adoption of dozens of new media types and technologies, many of them the result of convergences between older forms and the online and digital environments. These advances unsurprisingly have resulted in changes in the ways people work and spend their leisure time, particularly in the ways they seek information and diversion. As McGowan (2003) has so cogently put it, the actual spaces in which speech occurs have little significance removed from their cultural context but rather acquire their meaning from "the social expectations and understandings associated with different settings, the practices such spaces foster, and the relationships of those practices to First Amendment values" (p. 1563).

Broadcast and broadband television channels, to cite the most ubiquitous example, exist side-by-side on channel tuners in at least 80 percent of households, and car audio systems make broadcast radio, satellite radio, and recordings on CDs, audio cassettes, and MP3 players equally accessible options. Yet content on the broadcast radio and television channels are held to quite a different First Amendment standard because of their supposed unique pervasiveness.

As Chen (2005) has noted, while the U.S. Supreme Court has achieved "doctrinal stability" in its restrictions on obscenity since the *Miller* decision in 1973, its regulation of indecency in broadcast radio and television cannot be justified by its own strict scrutiny First Amendment standard, largely because its continued reliance upon the concept of conduit-based content restrictions cannot be justified. Moreover, its jurisprudence with respect to free speech rights is unlikely to improve, because it "has proved quite uneven in keeping pace with technological changes in communications" (p. 1360).

Perhaps so, though Justice David Souter demonstrated a certain adeptness in parsing the difference between the handling by public libraries of conventional acquisitions and online sources. In his dissent in *United States v. American Library Association* (2003), Souter observed that, "Whereas traditional scarcity of money and space require a library to make choices about what to acquire, blocking is the subject of a choice made after the money for Internet access has been spent or committed" (p. 236). Blocking, therefore, "is not necessitated by scarcity of either money or space" (p. 237).

For Chen (2005), the fundamental disconnect in the Court's approach to indecency doesn't require the proof of time to reveal itself; it lies in a comparison between two of the oldest communication technologies, radio and telephone. He observed with irony that the Court seems to believe "the most

cogent distinction between *Pacifica* and *Sable* lies in the nature of the two conduits: whereas broadcasting provides 'push' communications, the telephone is
the ultimate 'pull' tool" (p. 1434). But that distinction itself proves to be a blurry
one, since unwanted telephone solicitations are far more invasive than the act
of turning on a radio. He cited two more contemporary illustrations of the point:
cable may be more invasive than broadcast television, if only because the
sheer number of channels encourages scanning through dozens or hundreds of
channels where the unknown potentially awaits, and the internet seems more
invasive than any broadcast when one is "mousetrapped" by pop-up advertisements. He concluded that, "Pervasiveness does not divide, but rather unites all
forms of passive communication" (p. 1435).

## A Red Light in Cyberspace

In his analysis of the U.S. Supreme Court's rejection of the CDA in *Reno v.
ACLU* (1997), Craig (1998) observed that the CDA's blanket ban on indecency on the internet was not a form of "cyberzoning" as government attorneys
contended, because it was aimed, not at cleaning up the crime and other social
ills that might accompany offensive online content, but rather at the content
itself and thus was not in the spirit of the Court's *Renton* decision on TPM
restrictions. While his assessment no doubt was accurate, neither it nor *Reno*
precludes authentic forms of cyberzoning, particularly if they are established by
a non-state actor and are entirely voluntary.

Such was the scenario contemplated by the decision early in 2011 by
ICANN to authorize a new top-level domain suffix, .xxx, which could be used
by operators of websites specializing in adult content. The suffix originally was
proposed in 2004 by ICM Registry, one of many registrars ICANN contracts
with through a subsidiary to assign unique domain names, but it was initially
rejected by ICANN in 2007. ICM Registry appealed, and ICANN signaled in
2010 it was considering reversing its decision (Helft, 2011).

The analogy of this cyberzone as created by ICANN to a physical red
light district is as unmistakable as it is exhaustive: a clean, well-lighted place,
free of unwanted solicitations, viruses, and credit card thieves and home to businesses that "will operate by certain standards" (Helft, 2010b, p. B-1).
Establishments and their wares would be clearly labeled as "adult" and the whole
neighborhood would be easy for parents, schools, libraries, and offended adults
to block. Those who shouldn't be there or who do not wish to be there could
simply stay out.

To further the analogy, the process of creating the district—the zoning process—featured proponents (largely other businesses and their trade association) and opponents (mostly culturally conservative groups that see the district as an officially sanctioned den of iniquity and skeptics among the adult businesses.) According to ICM Chair Stuart Lawley, who, in his position, roughly is analogous to the president of the local Chamber of Commerce, more than 100,000 .xxx domain names had been preregistered by late June 2010, with another 400,000 expected in the following year. But a large number of potential registrants are resisting the tacit pressure to move their websites from .com domains to .xxx domains, apparently fearful that segregation might also facilitate eventual censorship of adult entertainment sites or make them irresistible targets for increased government regulation. While ICANN sees .xxx as defining a worry-free adult entertainment district some will avoid but others may feel safe to visit, Free Speech Coalition Executive Director Diane Duke, the spokesperson for the skeptics in this analogy, believes cyberzoning "will ghettoize our industry" (Helft, 2010b, p. B-1). The history of ICANN presents reason for both caution and assurance.

## The Chosen Instrument

The internet, the infamously vast interconnection of file servers that serves as the infrastructure for the web, was developed by the U.S. government in the 1960s. Arpanet, as it then was known, linked computers belonging to the Department of Defense and major research universities across the country. As the popularity of the internet spread with the widespread use of electronic mail and online bulletin boards, the Department of Commerce, which had assumed responsibility for assigning names and virtual addresses to the file servers that functioned as important nodes in the network, contracted with Network Solutions, Inc. in 1993 to perform that task (Weise, 1998). Difficulties with Network Solutions, Inc. and the desire to spread the ever-increasing workload among multiple private organizations led the Commerce Department to reassign the work to ICANN, which received its government charter in 1998, and its subsidiary, the Internet Assigned Numbers Authority (IANA). ICANN now effectively functions as the government's chosen instrument to develop and administer policies governing the use of the internet, while IANA's job is to coordinate assignment of domain names by a small group of private registry companies. Internet service providers and similar operators in turn assign user names within those domains (Bradbury, 2005).

Though the Clinton administration had pledged to grant ICANN greater autonomy, the U.S. government continues to oversee it through the Secretary of Commerce. The continuing specter of White House involvement in internet policymaking, abetted by the organization's somewhat secretive origins that contrasted sharply with the internet's traditions of consensus, openness, and collaboration (Weise, 1998), may explain why ICANN is perceived by some critics as being dominated by the United States, other Western governments, and multinational corporations (Oakes, 2004), despite a permanent governing board of 19 international directors (Weise, 1998) and an unblemished record of decisions made without interference from Washington (Wray, 2005).

Among the specific areas of concern have been fiscal management and free speech guarantees. Some critics have worried openly that financial controls built into ICANN's management plan are too vague, particularly since its board is composed primarily of appointees from the very groups that will be funding it. Others, including the Electronic Frontier Foundation (EFF), argued ICANN needed specifically to incorporate free expression into its bylaws, rather than to rely on the legacy of First Amendment protections that stemmed from its roots in the Commerce Department. As EFF spokesperson Alex Fowler put it, that reliance seems provincial for an organization that will be truly global in outlook and authority, since "the First Amendment of our Constitution that guarantees our citizens free speech is just a local ordinance in cyberspace" (Weise, 1998, p. 8-D).

## An Imperfect Solution

While being far from a panacea, creation of the .xxx domain suffix likely represents the best long-term solution to the problem of protecting underage and sensitive adult users from accessing indecent content online in that it appears to address a number of shortcomings, both constitutional and practical, of legislative approaches.

First, there are signs of, if not a paradigm shift in the approach to the control of offensive content in the media, at least a growing willingness to abandon the old one. Following a federal appeals court decision rejecting its policy on the fleeting use of expletives in live broadcasts in 2010, the FCC could again find its broader indecency policy under review by the U.S. Supreme Court, where evidence suggests at least some members may be growing increasingly skeptical of the unique pervasiveness argument that for over three decades has been at the foundation of the FCC's rationale for channeling indecency (Wyatt,

2010). If the Court is unwilling to permit TPM restrictions of non-obscene speech on broadcast channels, it is unlikely even the most narrowly crafted content restrictions on the internet could survive legal challenge, given the legacy of cyberspace as legally pristine territory unfettered by regulations on either content or structure. As Craig (1998) concluded, "Ultimately, indecent speech on the Internet may be one type of communication that simply cannot be legislated to protect the young" (p. 14).

Second, with the growing popularity of the web as a delivery vehicle for multiple forms of digital entertainment by commercial media interests, and with media companies more and more seeing social networking sites as vehicles for the distribution or marketing of entertainment content, the amount and variety of indecent content is likely to increase. International media conglomerates with the financial and political capital necessary to challenge and even circumvent policies and policy initiatives increasingly are willing to push the envelope when they perceive a market for edgy content. A report released in 2005 by the Center for Creative Voices in Media and Fordham University's Donald McGannon Communication Research Center connected the growing problem of indecency in radio, a medium dominated in large and medium markets by a small number of large station groups, to consolidation in media ownership, rather than to lax enforcement or a widening rift in social mores (Report links, 2005). As these behemoths, which once feared the internet in general and social media in particular as a threat to their intellectual property rights and as competitors for the leisure hours of their audiences, now embrace it, and as internet service providers attempt to divert user traffic to sites owned by those with whom they have financial relationships, it should come as no surprise if those sites also provide more challenging content. Myspace, which is owned by media giant News Corp., recently re-launched itself as a social media site focused on entertainment, signaling both a furthering of a trend toward specialization in social media and a joining of traditional entertainment forms with social networks as leisure activity. Even the author of the CDA, former Senator James Exon, had the foresight to admit, "I don't believe—and I don't think any reasonable person believes—that even a fully active, operating Communications Decency Act is going to clean up all pornography on the Internet" (Granatstein, 1996, p. 14).

Third, the .xxx suffix may restore some balance in the negotiation of access to offensive (and potentially inappropriate and even dangerous) content by underage internet users increasingly skilled at circumventing externally imposed age restrictions and by parents who, often unwittingly, enable them.

According to the Pew Research Center's Internet and American Life Project, even social networking sites such as Facebook, whose content largely may not be sexually explicit but which establishes age minimums to protect children from sexual predators and from unscrupulous marketers, may have a membership base of 12 years and under of as much as 38 percent. In one class at a San Francisco elementary school, half of the students had Facebook accounts, even though the site bars members under 13. According to a social media researcher at Microsoft, parents frequently help their children set up the accounts and tell them to lie about their age. That leaves them open to joining the networks of those they may think are friends or classmates but who in reality may be adults who will appropriate their images for use in child pornography (Richtel & Helft, 2011).

Fourth, cyberzoning would be an international solution befitting an industry sector and social concern that are both global as well. As Quittner (1997) noted in his account of the *Reno* decision, "Even if the government figures out a constitutional way to impose limited censorship on-line, these rules can apply only within the U.S.—and the Internet is international" (p. 29). The observation is particularly true of social media. By summer of 2010, the most pervasive of the social networking websites, Facebook, was reaching nearly half a billion users, 70 percent of them outside of the United States (Helft, 2010a). In addition, it avoids the technical difficulties associated with software filters and online ratings systems that are "beset with problems" (p. 29).

The imperfection in the cyberzoning solution arises primarily from the discretion websites have in adopting the .xxx suffix. If, as expected, 500,000 websites registered with ICM Registry by mid-2011, that still would be just 10 percent of the roughly five million to six million adult online sites thought to exist. In addition, many of those registrations could be defensive in nature, meaning owners had applied for them so that other applicants could not appropriate their names and brands (Helft, 2010b).

As has been noted, owners of some sites may harbor concerns that adoption of a .xxx domain suffix may amount to false security or be a ruse they may later regret. Others may favor the .com domain because their business model relies at least in part upon enticing those who might not seek out adult content but who might ultimately become interested (or whose personal information could be harvested as a valuable commodity.) But, indeed, quite apart from strategic misrepresentation and legitimate skepticism is genuine equivocation regarding how some sites might be correctly classified. The content of a social networking site such as Facebook, for example, by and large is bucolic if not completely innocuous; yet, dialogue represented by postings on walls of users and groups may contain explicit content, as may photo albums and videos posted by the owners. Likewise, fan tribute sites celebrating the works of artists such as Tom of Finland, Chris Rock, or Sarah Silverman might contain language or images many would find inappropriate for minors even though a .xxx domain would hardly characterize their content as a whole.

## Conclusion

The adoption of .xxx domains represents a bold experiment in industry-wide self-regulation by ICANN, the body designated to coordinate, to the extent that anyone can, information exchange on the web. It also gives form and substance to the concept of cyberzoning in the form of an analogous system that functions to segregate content in virtual space, much as zoning laws segregate behavior and interchange in physical space.

Its ultimate effectiveness as an alternative approach to channeling expression that is both protected and challenging without the risks of infringing upon the right both of some to speak and others to receive will be determined not just by the more mechanical considerations represented by the installation and maintenance of software to block access to .xxx sites and by the good-faith use of them by sites, but by the larger and more elusive consideration of cultural perceptions of them.

Similar self-regulatory regimes that effectively brand content and then rely upon the action (or even attention) of the audience arguably have had varied outcomes.

Ratings given television programs in the United States as mandated by provisions of the Telecommunications Act of 1996 provide specific information concerning the age-appropriate content of programs but also the nature of the content that necessitates the rating. One program may, for instance, be appropriate for teens but inappropriate for younger children because of its level of violence. Another may be appropriate only for adults because of language. The program ratings are both ubiquitous and lightly regarded (Bushman & Cantor, 2003). Most viewers may be peripherally aware of them as they appear on screens at the beginning of program segments but few understand them (Kaiser Family Foundation, 1998) or act upon them by using them to guide family viewing (Signorelli, 2005). Ratings given films, on the other hand, provide fewer specifics but have had far greater cultural impact, though the impact at times has been unintended. Most exhibitors appear to be vigilant regarding the enforcement of the age restriction on "R" and "NC-17" films (Septimus, 1996). Comparatively few films are released unrated. The system enjoys widespread cultural meaning and currency. One might speak of a "G-rated" vacation or an "R-rated" conversation secure in the knowledge the audience understands the constructs. As apparently effective as the system has been, it has its seeming dysfunctions as well. While the meaning of "X-rated" also is well-understood, the failure of the Motion Picture Association of America (MPAA) to trademark it in a timely manner after its introduction led to its re-appropriation by the adult-film industry (whose producers typically are not MPAA members) and ultimately to its abandonment by the MPAA (Septimus, 1996). In addition, there is evidence many studio executives strategically edit films to earn particular ratings that are thought to target particular audiences (Attanasio, 1985). "G-rated" films are thought to repel adult audiences. "R-rated" films suggest to teens more mature content and are more attractive (Cantor, 1998); in addition, the need for accompaniment by a parent may lead to another ticket sale.

The relative success or failure of cyberzoning also likely will depend upon its cultural interpretations. The domain identifiers ".com," ".net," ".org," and ".gov" probably are understood, at least to some extent, by many regular internet users; newer and more specific suffixes likely are not as well understood. To the extent that nation-specific suffixes such as ".co.it" are understood, it may be by those with friends or business contacts overseas or, more pejoratively, through encounters with nuisance e-mails, worms, and viruses.

If the denotative meaning of ".xxx" becomes culturally dominant, it may succeed at its intended purpose. If unattractive or less accurate connotations attach themselves to it, reluctance to adopt it by sites for which it was intended or misdirected traffic by users will render it a failure.

## References

*ACLU v. Gonzales,*478F. Supp. 2d 775 (E. Dist. PA, 2007).

*ACLU v. Reno,* 929 F. Supp. 824 (E. Dist. PA, 1996).

*Ashcroft v. ACLU,* 542 U.S. 656 (2004).

Associated Press. (2010, June 27). Porn sites closer to .xxx web address. *Cincinnati Enquirer,* p. G5.

Attanasio, P. (1985, 17 June). Desperately seeking standards: The rating game. *New Republic,* X, Vol 192, 16–18.

Bradbury, D. (2005, 29 November). Who owns and administers the internet's addresses? *Computer Weekly,* X, 52.

Bushman, B. J., & Cantor, J. (2003). Media ratings for violence and sex: Implications for policymakers and parents. *American Psychologist, 58*(2), 130–141.

*Butler v. Michigan*, 352 U.S. 380 (1957).

Cantor, J. (1998). Ratings for program content: The role of research findings. *Annals of the American Academy of Political and Social Science, 557*, 54–69.

Chen, J. (2005). Conduit-based regulation of speech. *Duke Law Journal, 54*(6), 1359–1456.

Child Online Protection Act (COPA). 47 U.S.C. § 231 (1998).

Children's Internet Protection Act (CIPA). Pub. L. 106–554 (2001).

*City of Renton v. Playtime Theatres*, Inc. 475 U.S. 41 (1986).

Craig, J. R. (1998). *Reno v. ACLU*: The First Amendment, electronic media, and the internet indecency issue. *Communications & the Law 20*(2), 1–15.

*Cruz v. Ferre*, 755 F.2d 1415, 1416 (11th Cir. 1985).

*Denver Area Educational Telcommunications Consortium v. FCC*, 518 U.S. 727 (1996).

Docter, S. (1992). An alternative justification for regulating broadcast indecency. *Journal of Broadcasting & Electronic Media, 36*(2), 245–248.

*Dunlop v. U.S.*, 165 U.S. 486 (1897).

Facebook. (2011). Terms of service. Retrieved from http://www.facebook.com/help/search/?q=terms+of+service

*Ginsberg v. New York*, 390 U.S. 629, 634 (1968).

Google. (2007). Terms of service (archived). Retrieved from http://www. google.com/intl/en/policies/terms/archive/20070416/

Google. (2012). Terms of service (revised). Retrieved from http://www. google.com/intl/en/policies/terms/

Granatstein, L. (1996, April 8). A battle over bytes: Should cyberspace be censored? *Time*, 14.

Helft, M. (2010a, July 8). Facebook makes headway around the world. *New York Times*, p. B1.

Helft, M. (2010b, June 26). For X-rated, a domain of their own. *New York Times*, p. B-1.

Helft, M. (2011, March 19). Pornography sites will be allowed to use .XXX addresses. *New York Times*, p. B-5.

Kaiser Family Foundation. (1998). *Parents, children and the television ratings system: Two Kaiser Family Foundation surveys*. Menlo Park, CA: Author.

LinkedIn.(2011). User agreement. Retrieved from http://www.linkedin.com/static?key=user_agreement&trk=hb_ft_userag

McGowan, D. (2003). From social friction to social meaning: What expressive uses of code tell us about free speech. *Ohio St. L.J., 64*, 1515.

*Miller v. California*, 413 U.S. 15 (1973).

Oakes, C. (2004, March 8). External forces chip away at Internet's overseer. *International Herald Tribune*, p. 13.

Quittner, J. (1997, July 7). Unshackling net speech. *Time*, Vol 150, 29.

*Regina v. Hicklin*, L.R. 3 Q.B. 360 (1868).

*Reno et al. v. ACLU et al.*, 521 U.S. 844 (1997).

Report links consolidation of broadcast ownership to rise in indecency complaints. (2005). *Media Report to Women, 33*(4), 3–4.

Richtel, M., & Helft, M. (2011, March 12). Facebook users who are under age raise concerns. *New York Times*, p. B-1.

Rivera-Sanchez, M. (1994). Developing an indecency standard. *Journalism History, 20*(1), 3–15.

*Roth v. United States*, 354 U.S. 476 (1957).

*Rust v. Sullivan*, 500 U.S. 173 (1991).

*Sable Communications of Cal., Inc. v. FCC*, 492 U.S. 115 (1989).

Septimus, J. (1996). The MPAA ratings system: A regime of private censorship and cultural manipulation. *Columbia-VLA Journal of Law & The Arts, 21*(1), 69–93.

Signorelli, N. (2005). Age-based ratings, content designations, and television content: Is there a problem? *Mass Communication & Society, 8*(4), 277–298.

Telecommunication Act of 1996,110 Stat. 56 (1996).

Twitter. (2011). Rules. Retrieved from http://support.twitter.com/articles/18311-the-twitter-rules#

*United States v. American Library Association*, 539 U.S. 194 (2003).

*United States et al. v. Playboy Entertainment Group, Inc.*, 529 U.S. 803, 2000.

Weise, E. (1998, October 7). Net name control hits a snag; disputes hinge on who will set rules. *USA Today*, p. 8-D.

Wray, R. (2005, October 12). EU says internet could fall apart; developing countries demand share of control; US says urge to censor underlies calls for reform.*The Guardian*, p. 27.

Wyatt, E. (2010, July 14). FCC indecency policy rejected on appeal. *New York Times*, p. B-1.

# · 1 0 ·

# Social Media, Public Relations and Ethics

SUZANNE BERMAN

## Introduction

Public relations professionals have been among the first to realize the over-whelming power of social media. Recognizing that both public relations and social media are about building relationships, public relations specialists were early to understand that social media could enable them to connect with the public in a way they had never been able to do before. Today, they see how profoundly the Internet, and social media in particular, are changing the face of the industry.

Social media is being utilized on an ever-increasing basis by corporations and other organizations. More than two thirds (69%) of the current Fortune 2000 companies are using social networking sites (McCorkindale, 2010) and of those not yet fully engaged, almost all organizations are exploring ways to increase their social media communication efforts and measure its effectiveness.

During the past few years, significant change has occurred in the way that public relations activities are conducted. Recently, there has been a shift away from many of the more traditional methods and techniques utilized in public relations campaigns and a move toward online strategies that relate more to the way people communicate today. The press release, although still pervasive, has been pushed aside somewhat to make room for the addition of the social media release (SMR) with its embedded multimedia elements and easier, more direct distribution methods. Public relations people are now publishers in their own right, distributing stories directly to the public through a variety of media tools and social media platforms. Social networking sites like Twitter and Facebook have become meaningful places for public relations professionals and the companies they represent to connect with consumers and journalists. Social media has enabled public relations professionals to locate specific audience groups, listen to what they are saying, and target them in very specific ways. Overall, digital media has enhanced the ability of public relations professionals to engage in personal relationships directly with a variety of publics, connecting social media and public relations ever more closely.

But with these numerous benefits come questions about whether public relations experts are utilizing social media ethically. Over the years, the term *public relations* has come to mean different things to different people. Often negative connotations had been associated with the profession. Sometimes words such as spinning or lying would be attached to it, rather than people seeing public relations as the serious profession it has become for many companies, agencies, and nonprofit organizations. Unfairly dubbed or not, the ethics that a public relations professional abides by is crucial to the profession's credibility. With the Internet, and the limited amount of regulation governing this new frontier, the role of ethics in public relations has become even more critical.

## What Is Social Media Public Relations?

Social media describes the online technologies and practices that people use to share content, opinions, insights, experiences, perspectives and media itself (Wilcox & Cameron, 2010). Utilizing these technologies has become an important strategy in public relations campaigns. Social networks, apps, video and photo sharing sites, news aggregators, social book marking and other features of the Internet have enhanced collaboration and sharing between companies and their publics. Entire public relations campaigns are now built with these platforms in mind. It is no longer sufficient to announce, release or simply broadcast information about a company or product. Once people are aware of the product, a new dynamic kicks in: people learn from each other. Social technologies have revved up that word-of-mouth dynamic, increasing the influence of regular people while diluting the value of regular marketing (Li & Bernoff, 2008).

According to Brian Solis and Deidre Breakenridge (2010) social media and Web 2.0 are altering the entire media landscape, placing the power of influence in the hands of regular people with expertise, opinions and the drive and passion to share those opinions. This people-powered content evolution augments instead of replaces traditional media and expert influence. And, in the process, entirely new layers of top-down and bottom-up influence have been created. These layers dramatically expand the number of information channels (one-to-one, one-to-many and many-to-many; Solis & Breakenridge, 2010). Social media public relations is predicated upon the idea that there are communities of people that can be reached directly and influenced through a mixture of conversation and listening. Based on this, social media public relations efforts focus more on a two-way method of interaction. In this new world, companies augment and let go of the push broadcast mechanisms associated with traditional marketing and message control, enabling customers to internalize information and, in turn, share their reaction and interpretations (Solis &Breakenridge, 2010).

## Elements of Social Media Public Relations

Just as radio and television required public relations professionals to adapt their activities to meet the needs of those mediums in the middle of the last century, the World Wide Web (web) has created a whole new set of activities for the public relations field today. The fast growth of Internet use, and how that has impacted the way people receive information, means that public relations activities had to change. Companies realize that the Internet has altered the way people get information from them. Communicating through the Internet means that people or members of an organization are pulling information off the Internet about an organization rather than the organization pushing information onto them as in the case of the regular new release dissemination. Making all appropriate information available on websites, along with accompanying links, means that customers, media reporters, investors, financial analysts, employees, government officials, regulators, activists and others, can get whatever information they want and put it together in any order they want, without going through the media or public relations professional (Lattimore, Baskin, Heiman, &Toth, 2009).

According to a study conducted by Wright and Hinson (2008), there is considerable agreement suggesting blogs and social media have enhanced public relations practice. Two-thirds of the study's respondents (66%) believe social media has enhanced public relations and 60% feel the same way about blogs. Eighty-nine percent of those surveyed think blogs and social media influence news coverage in the traditional media (newspapers, magazines, radio and television), while 72% say the reverse also is true. There is very solid agreement (84%) that blogs and social media have made communications more instantaneous because they force organizations to respond more quickly to criticism (Wright & Hinson, 2009).

In perhaps one of the largest studies of corporations and blogs (Backbone Media, 2005), respondents perceived the highest values attained from blogging to the organization to be quick publishing, thought leadership, community building, sales and online public relations. These attributes continue to play an important role in the effectiveness of blogging today.

Responding to these trends, and the growing belief that online communications can have a powerful impact on changing the value and scope of the public relations industry, several new or revised public relations activities have arisen.

## The Social Media Press Release

The press release was invented in 1906 by one of the fathers of modern-day public relations, Ivy Lee. Since that time, the press release has been widely used by organizations as a way to disseminate information in an open and quick manner through the media. Today, with the advent of digital media, the press release is getting a considerable makeover, getting retrofitted to meet the needs of a multimedia hungry audience. What started as a tool for the media only has evolved into a tool designed to reach customers directly through search engines and news aggregation services. Distribution of news releases via a wire service such as PRNewswire offers additional value in the form of SEO (search engine optimization). Integrating keywords, phrases and embedded links optimizes their "findability" and rank within traditional search engines such as Google and Yahoo (Solis and Breakenridge, 2010). Many organizations feel that if they are not in the first two pages of search results, their message is being lost.

Increasingly, the traditional release is being replaced by the SMR. Essentially, the difference between a traditional press release and a SMR focuses on the intended audience. A traditional press release targets journalists in hopes of interesting them in writing a story. Traditional releases are written to give journalists facts and information that allow them to craft their own story angle any way they like, using the information provided. The SMR, on the other hand, is intended for customers, bloggers and the public at large, and it is designed to encourage conversation and sharing. The SMR allows companies and organizations to give information directly to the public rather than having to go through the media. Since the SMR is meant to go directly to audiences, the style of the SMR is quite different from the traditional release and allows companies to develop their own story angles and to encourage sharing by including social media tools.

A SMR can be broadly defined as a single page of web content designed to enable the content to be removed and used on blogs, wikis and other social channels. Although SEO has certainly broadened the reach and scope of the traditional release, the SMR has taken the traditional press release even one step beyond SEO by combining news facts and social assets into one easy-to-digest and improved tool. A SMR can be broadly defined as a single page of web content designed to enable the content to be removed and used on blogs, wikis and other social channels. In practice, SMRs feature multiple embedded links (YouTube video, Flickr slideshow, SlideShare presentation, etc.) and blocks of text similar to those found in traditional releases (spokesperson quotes, boilerplate and contact information; Capstick, 2010).

Today the SMR continues to evolve, led by the International Association of Business Communicators (IABC), which has assumed responsibility for coordinating efforts to develop standards for the SMR, combining flexible formatting options with a tagging standard. The goal is to make business news usable by online reporters and bloggers, as well as more discoverable in search engines. The desire to retrofit the traditional press release began six years ago with a blog post by Tom Foremski titled, "Die, Press Release. Die! Die! Die!" (Foremski, 2006) where Foremski's post called for companies to parse the information in their press releases so he could pick the elements he wanted to form his story. The traditional release, written in narrative, makes that hard to do. Hearing the frustrations of bloggers like Tom, Todd Defren at SHIFT Communications released the first template for a SMR, and since then, common standards are being worked on continually(IABC, 2008). Enabling journalists and bloggers alike, along with the public, to produce online stories quickly and easily has provided SMRs with enormous value in a digitally focused world. They have gone a long way toward ensuring that the stories and messages PR professionals want to disseminate are being heard through a variety of different media platforms.

## Posting Viral Videos

Like advertisements, creating videos and posting to video-sharing sites such as YouTube and Vimeo are other ways companies raise visibility, and these methods have fast become an important part of the public relations toolbox. Videos can range from product demos, to speaker presentations to fun, creative videos targeted at consumers. As public relations professionals are always looking for ways to highlight a client's message to journalists as well as consumers in hopes of creating a strong following and appreciation for the brand, viral videos have become a vital part of this social networking process.

While video news releases (VNRs) have been around for a long time, viral videos have come of age with social media. With their built-in capacity for sharing, viral videos have become increasingly popular. Much like the traditional press release, the VNR is effective in distributing stories and information for the television market in a concisely packaged format. Although effective, they are extremely expensive and difficult to produce and lack the interactive component that viral videos have. Now with the advent of social media, videos, and in particular viral videos, are quickly becoming the most important public relations tool around today. According to the Pew Internet & American Life Project, over half of all Internet users frequent a video-sharing website (Madden, 2009). Online video has become a bigger fixture of everyday life, gaining 19% of all Internet users who use video-sharing sites during a typical day (Madden, 2009). The public is increasingly searching out video-based information, and companies that recognize this growth are producing video programming to speak to these audiences. By picking up a camera and speaking directly to audiences, companies are able to tell stories to people who enjoy visual media, and more importantly, they are able to get those videos seen over and over again through video-sharing sites. Companies are now creating channels on YouTube, tagging each video with keywords to help people find them, and creating embedded codes that enable people to share the videos with other. Companies such as BlogTV are helping organizations produce episodic broadcasting, and technology such as mobilecasting is allowing video to be streamed live from a mobile phone. Social media has enhanced the powerful role that video technology can play in the public relations mix.

## Engaging in Social Networking

The participatory nature of social media has increased its usage and popularity, making it America's favorite online activity (Mui & Whoriskey, 2010). Evidence of this can be seen when social media became a regular part of the political environment for voters in the 2010 midyear election in the United States (Smith, 2011). With so many people using social networking sites such as Facebook and Twitter, not engaging in social networking is simply no longer an option in the practice of public relations. Companies develop Facebook pages; they use Twitter, which enables them to tweet directly to customers. Since social networking is about engaging in conversations, customers leave comments and expect quick responses directly from companies. This constant online dialogue has also enabled public relations professionals to monitor what is being said about their clients and their brands, using easy access monitoring tools like Google Alerts. By proactively managing one's online presence, companies ensure that reliable and accurate information is always available, and they no longer have to rely on third parties to make this happen.

Most companies in the Fortune 50 are taking advantage of the opportunities of Facebook, (McCorkindale, 2010). Some sites, such as Ford's, have a strong fan base, and others, such as Dell and Microsoft, respond to customer issues and offer product deals on their Facebook pages. But, these companies are among the most active, while others merely post news releases and mission statements. Most companies are not using their Facebook pages sufficiently and are not disseminating updates and information about the organization, especially in terms of corporate social responsibility, frequently enough (McKorkindale, 2010).

When used effectively, however, social networking can encourage a viral buzz around a compelling story, news or product. By launching a social community through Facebook and other social networking sites, companies provide a platform to share photos, videos, news and other information to engage users online and drive word of mouth. There is still much more to be done in this area, as public relations professionals become accustomed to being content generators.

## Bloggers and Blogging

Edelman and Intelliseek (2005) described blogs as being "easily published, personal web sites that serve as sources of commentary, opinion and uncensored, unfiltered sources of information on a variety of topics" (p. 4). Robert J. Key (2005) suggested that many weblogs began sporadically as vanity publishing because "anyone with an opinion about anything could create, in a matter of minutes, his or her own web site for publishing news, opinion, commentary and links to other sites" (p. 18).

Whether or not they are viewed as credible sources, public relations professionals find themselves both having to court bloggers, much like they do journalists, and create and write blogs on behalf of clients or companies. Blogs are useful tools that allow companies to engage in dialogues directly with the public. With blogs have come new influencers, adding to the voices that have helped guide consumers and decision makers. If anything, these new influencers are challenging the very foundation of public relations, forcing professionals to revisit, and often repair, the process of how news and information is shared (Solis & Breakenridge, 2010). It has therefore become imperative for public relations professionals to interact with this important community.

From 2005 to today, the blogosphere has exploded (Smith, 2011). Many bloggers consider themselves commentators on today's news and, in fact, are likely to be part of the audience for which they are writing (Lattimore et al., 2009). For this reason, public relations professionals are adopting new ways to approach this audience, ways that differ from many of the methods employed with traditional media relations, often approaching bloggers as they would anyone in a social situation by introducing themselves and getting to know them before expecting a story or posting. There are many examples of how companies are using bloggers to successfully reach customers and help gain awareness. The reach and influence of many bloggers, such as the mommy blogging community, can have a startling impact on public relations efforts. According to a 2009 study by BlogHer, ivillage and Compass Partner, 23 million women read, write or comment on blogs weekly (Mendelsohn, 2010). Certain bloggers in every market have the sheer numbers of followers behind them to not only influence but also to drive reporters in traditional media to cover the same topics (Solis and Breckenridge, 2010).

In addition, many companies now engage in blogging of their own. According to an article in Mashables mall businesses with corporate blogs receive 55% more traffic than small businesses that don't blog (Swallow, 2010). Many companies use their blogs as platforms to share their insights and opinions, but today companies are also using their blogs as vehicles that invite the public to share their comments and concerns. For example, HP has many blogs, as opposed to some companies that choose to have one corporate blog, like GM's Fastlane. HP now has 50 blogs. While the blogs generate traffic, awareness is not the most important benefit. The benefit is that HP can respond to its customers with frequent and diverse dialogue that is updated as bloggers hear from customers or from comments on the blog (Li &Bernoff, 2008).

## Creating Communities

In addition to building communities through social network sites like Facebook and Twitter, companies are also developing their own websites, soliciting involvement from the public and building relationships directly. Proctor and

Gamble (P&G), for example, in an effort to connect with the hard to reach feminine care market, was looking for a way to engage young, new potential customers without embarrassing them through blatant advertising. The company decided to help solve young girls' problems through a social community entitled Beinggirl.com, a site about issues that young girls deal with on a daily basis. To become part of the dialogue among young girls, P&G created a social network. Because its aim was to solve customers' problems instead of its own, and talk about issues relevant to the target audience, the customers were willing to share and the community has grown. Combining subtle brand messages and free samples enabled P&G to become part of the dialogue about issues of relevance to their tampon-using customers (Li & Bernoff, 2008).

## The Role of Ethics in Public Relations

The role of ethics in public relations is a complex one, with professionals constantly having to balance the needs of the public interest with the needs of their clients, all the while adhering to a clearly defined code of professional ethics, as well as their own sense of right and wrong. Professional organizations such as the Public Relations Society of America (PRSA) and the IABC have developed sets of ethical standards that have been instrumental in advancing professionalism in the field. PRSA has a comprehensive code of ethics and believes that "values are vital to the integrity of the profession" (PRSA, 2012). Its values are as follows:

> **Advocacy:** Serving the public interest by acting as responsible advocates for clients or employers; **Honesty:** Adhering to the highest standards of accuracy and truth in advancing the interests of clients and employers; **Expertise:** Advancing the profession through continued professional development, education and research; **Independence:** Providing objective counsel and being accountable for individual actions; **Loyalty:** Being faithful to clients and employers, but also honoring an obligation to serve the public interest; **Fairness:** Respecting all opinions and supporting the right of free expression. (PRSA, 2012)

While these ethical standards have been helpful for establishing codes for the overall practice of public relations, many professionals believe that specific situations and issues require more defined standards that address concerns that have arisen as a result of the seismic shift that has occurred in the way people communicate today. Public relations practitioners are using the Internet and online services to send and find breaking news and feature stories, speeches, photos and information. Although easily accessible, many of these resources are copyrighted and are sometimes used without permission. Copyright issues are sometimes overlooked in a world that is not only built but evaluated by its ability to share content. Additionally, conducting public relations in a digital environment allows for a sense of anonymity that enables public relations professionals to create online interest in products and services, sometimes without revealing who they are working with and oftentimes resulting in deception or misrepresentation. As a result of these, and similar activities, the Arthur W. Paige Society, an organization of senior-level communication executives, and 10 other major public relations organizations, decided it was time to establish a set of principles for public relations on the web (Wilcox & Cameron, 2010). These principles follow:

~ Disclose any affiliations in chatroom postings
~ Offer opportunities for dialogue and interaction with experts
~ Reveal the background of experts, disclosing any potential conflicts of interest or anonymous economic support of content
~ Practice principled leadership in the digital world, adhering to the highest standards. (Wilcox and Cameron, 2010)

Additionally, in 2008, the Federal Trade Commission, recognizing the public's growing dependence on blogs, instituted a set of guidelines that require bloggers, prominent tweeters and people commenting on Facebook, to disclose any paid endorsements to their followers, online friends and readers (Gross, 2009).

While guidelines and principles have been helpful as a start to understanding how to conduct public relations in an online world, each day new scenarios come up that require public relations people to reflect upon the way they conduct their business in today's digital environment.

## Social Media and Public Relations Ethical Dilemmas

With each new tool, social media ethical dilemmas continue to arise. Here are some examples.

### Tweeting

Today, many companies have Twitter accounts that are written by industry executives who enjoy a significant following. Keeping up with constant tweets can oftentimes be challenging for busy executives. Sometimes, public relations professionals are asked to fill in and write the tweets on behalf of the company and the executive. This kind of request is typically not motivated by a desire to manipulate but rather by the executive's many other time commitments and busy schedule. According to social media expert Todd Defern (2010), the executive's commitment to online engagement is so fierce, he or she doesn't want to abandon it even for an important event. In this case, public relations professionals are put into a difficult position of either saying "no" to a client or knowingly misleading the public. According to public relations experts, an appropriate and ethical solution is to agree to tweet on behalf of the client but insisting that the executive tweet as well, explaining when someone is filling in (Defern, 2010)

## Ghostblogging

Whether referred to as citizen journalism or blogging there, is little doubt that there are many voices on the Internet all clamoring to be heard. According to Technorati (2010) an online directory for blogs and their rankings, more than 112 million blogs exist and cover everything from business, sports, lifestyles, fashion and entertainment. Blogs offer valuable information, and therefore, they have become important influencers. Today's public relations professionals incorporate blogs into their campaigns, and in many cases, they create highly defined blogger relations programs in order to ensure that these influential voices are constantly updated about their client or company.

But with this rapid growth of the blogosphere has come the need for more stringent standards. While some blogs maintain high standards, there are some that are inaccurate, biased, or poorly written. Without a body governing the blogosphere, many violations occur. Scholars have begun researching how people view mainstream media versus blogs and have found that traditional news media receive higher scores than blogs and social media in terms of accuracy, credibility, truthfulness and ethics, and more people consider traditional news media to be honest, truthful and ethical, while not holding these same expectations for blogs and other social media (Wright and Hinson, 2009).

Perhaps one of the most serious ethical dilemmas in terms of blogging is that of ghostblogging and disclosure. The issue of ghostblogging, a new version of ghostwriting, in which the writer is not fully identified, is of ongoing concern on the Internet. The issue of disclosure as it relates to blogs first came to light in 2006, when it was revealed that Walmart's blog, "Wal-marting Across America," was in fact not written by independent supporters of Walmart but rather by people paid by Walmart's public relations firm, Edelman, to travel across America and report on their dealings with Walmart stores. The Walmart blog began on September 27, 2006, featuring the journey of Laura and Jim, a couple on their maiden trip in a recreational vehicle, capturing lives and stories as they journeyed from Las Vegas to Georgia and parking for free at Walmart stores (Gogol, 2006). This fake blog became the beacon for unethical behavior in the blogosphere and ushered in a renewed emphasis on transparency. Despite the Walmart incident, blogging continues to be a powerful public relations tool, and professionals continue to explore the best ways to blog in an open and transparent way.

There are two types of blogs, the corporate blog and the personal blog. For the successful corporate blogger, content becomes part-and-parcel of an overarching communications strategy (Defern, 2010). With the need to constantly update content on corporate blogs, outside public relations consultants are hired to ghostblog, raising the question of whether ghostblogging is ethical. Many public relations professionals believe that while ghostblogging for a personal blog would be unethical, ghostblogging for a corporate blog is no different than drafting an article for a company newsletter.

## Commenting on Blogs

Similarly, on occasion, public relations agencies are asked to help clients engage bloggers by commenting on industry blogs. The goal is to insert the client's executives and perspectives into industry conversations, to help them improve credibility and ultimately to create valuable relationships with influencers (Defern, 2010). With hundreds of blogs to monitor and engage with, companies are looking for outside help to maintain a high level of engagement with the blogging community. Some agencies disclose who they are, while others are asked to comment as though they were the client. While the former approach is understandable and does not misrepresent the client, the latter does raise some interesting ethical dilemmas. Monitoring blogs for clients is a valuable service that is provided by public relations agencies. They make sure clients are well-informed about trends and opinions, and by commenting regularly, they ensure that influential bloggers understand how much the clients care about bloggers' content. So how does one weigh these benefits with the ethical dilemmas inherent in the activity of ghostblogging? For many, the solution is considered to be full disclosure. As long as the public is aware of who is commenting, then a little outside help is acceptable.

Perhaps one of the most blatant violations of disclosure is the case of Whole Foods Market Chief Executive Officer John Mackey, who had been anonymously posting to a Yahoo! stock message board of competitor Wild Oats in what appeared to be an attempt to drive Wild Oats' stock down in advance of the Whole Foods acquisition (Stone & Richtel, 2007).

## Employee Blogging

The majority of employee blogs fall into two categories: blogs written by employees intended for other employees and blogs written by employees intended for external audiences, such as customers or potential customers. The common thread for employee blogs, regardless of the intended reader, is that the employer has most likely little to no oversight or control over the content, especially in the case of blogs created on employees' own time and equipment. For any blog to be credible, it should deal with both positive and negative comments and be transparent (Edelman, 2005).Many companies are a bit hesitant when it comes to employee blogs, because an employee may disclose confidential business information, talk negatively of other employees or of the company and display an interest that is in conflict with business policies. There is a thin line between the company's best interests and the employees' First Amendment right to free speech (Black, 2010). According to Wright and Hinson's 2006 study, close to half of the respondents they asked said it was ethical to ask employees to refrain from posting negative comments on their blogs, while 2007 results showed agreement dropping all the way to 29%, and in 2008, that agreement slipped to 25%. Survey respondents continue to disapprove more each year when asked if it is ethical for organizations to conduct research about or monitor information that their employees are communicating via blogs and other social media (Wright and Hinson, 2008).While this drop is not dramatic, it indicates a clear signal that employees, and the public in general, in this social media era, believe that they should be free to say whatever they like online, and it brings into question the individual's right of freedom of speech.

## Engaging with Fans Online

Asking outside public relations consultants to engage with fans online poses some of the same kinds of concerns as that of ghostblogging and commenting on blogs. Moderating a client's Facebook fan page and engaging with fans online through a variety of different social networking methods have become time-consuming endeavors. Many companies have only one staff member dedicated to this and oftentimes hire outside agencies to assist. These agencies do not typically proactively identify themselves, but when the situation calls for it, agencies often reveal that they are representatives of the company. Whether or not the public believes they are interacting with specific individuals rests solely on their own due diligence to determine. While not deliberately misleading, this practice does raise concerns about transparency and responsibility.

## Information Sharing

With an increased reliance on social networking sites has come the proliferation of user data and information sharing. Because of this, privacy continues to be an important issue surrounding public relations and its use of customer information. Public relations planning often involves extensive research into the behaviors of target audience groups, and as such, the market for personal data about Internet users has grown exponentially with the growth of social media. Out of this growth a practice known as "scraping" has arisen. With scraping, firms offer to harvest online conversations and collect personal details from social networking sites, resume sites, and online forums where people might discuss their lives (Angwin & Stecklow, 2010. While some companies collect personal information such as e-mail addresses, phone numbers and posts on social network sites, other companies offer listening services that monitor news sources, blogs and websites to see what people are saying. This burgeoning industry has raised a number of questions about the changing definitions of public and private in the Internet environment and what people can expect about what will happen with the information they share online. While there have been federally mandated rules to ensure the privacy of individuals, these rules do not extend to the social media arena (Grier, Thomas, & Nicol, 2010).

Increased concern regarding user security and privacy on social networking sites underscores the need for more clarity surrounding the different types of user data. Schneier (2010) offered a taxonomy of social networking data that includes *service data, disclosed data, entrusted data, incidental data, behavioral data* and *derived data*. *Service data* is the data you give to a social networking site in order to use it. Such data might include your legal name, your age, and your credit card number. *Disclosed data* is what you post on your own pages: blog entries, photographs, messages, comments and so on. *Entrusted data* is what you post on other people's pages. It's basically the same information as disclosed data, but the difference is that you don't have control over the data once you post it—another user does.

*Incidental data* is what other people post about you: a paragraph about you that someone else writes or a picture of you that someone else takes and posts. Again, it's basically the same information as disclosed data, but the difference is that you don't have control over it, and you didn't create it in the first place. *Behavioral data* is data the site collects about your habits by recording what you do and whom you do it with. It might include games you play, topics you write about, news articles you access (and what that says about your political leanings) and so on. *Derived data* is data about you that is derived from all the other data. For example, if 80% of your friends self-identify as gay, you're likely gay yourself (Schneier, 2010).

## Conclusion

The rise of online communication, and with it social media, has enhanced the reach and value of the public relations field. Wright and Hinson's 2008 study indicated that public relations practitioners across the board feel that social media has significantly influenced the practice of public relations and found that social media has enhanced the two-way communication flow between a company and its publics. For the most part, these developments have been positive, but out of this growing interconnectedness, and the amount of information that it creates, questions have arisen that are forcing the industry to evaluate social policy issues, including electronic privacy, security, commerce and civil rights.

The digital environment has made it easier to gather and transmit information. This has had enormous value to both corporations and individuals, but there are risks surrounding authenticity, transparency and privacy. These risks continue to mount, as new communication developments surrounding GPS (Global Positioning System) technology are being explored and information-sharing capabilities are enhanced. As public relations professionals become publishers, allowing them to disseminate information directly to the public without first transmitting this information through a gatekeeper, their accountability and level of responsibility needs to increase. Web 2.0 technology has facilitated collaboration and sharing among users with such social methods as wikis, social networking sites, apps, video sharing sites, virtual reality sites, news aggregators, photo sharing sites, and mobile technology.

With new technologies developing at record speed, public relations profes-sionals are learning to balance the many inherent benefits stemming from these technologies with an ethical and moral code to guide them as they develop campaigns that are respectful of individual rights and freedoms. Ethical codes of conduct currently being developed and reviewed on an ongoing basis by public relations member organizations are serving an important role in pro-viding the guidance to ensure that misrepresentation and deliberate duplicitous work on behalf of public relations professionals will not be tolerated. With open and transparent use of social media, the field of public relations is uniquely poised to grow in the Internet era.

## References

Anguin, J., & Stecklowe, S. ( 2010, October 11). Scrapers dig deep for data on web. Retrieved from http://online.wsj.com/article/SB10001424052748703358504575544381288117888.html

Backbone Media.(2005). Corporate blogging survey. Retrieved from www.backbonemedia.com/blogsurvey

Black, T. (2010, April 26). How to handle employee blogging. *Inc.* Retrieved from http://www.inc.com/guides/2010/04/employee-blogging-policy.html

Capstick, I. (2010, April 23). Social media release must evolve to replace press release. Retrieved from http://www.pbs.org/mediashift/2010/04/social-media-release-must-evolve-to-replace-press-release113.html

Defern, T. (2010, January 5). Tweeting under false circumstances: Social media ethical dilem-mas. *PR Squared.* Retrieved fromhttp://www.pr-squared.com/index.php/2010/01/tweeting-under-false-circumstances-social-media-ethical-dilemmas

Edelman. (2005, July). Managing employee bloggers. Retrieved from http://old.edelman.com/image/insights/content/Edelman%20Employee%20Thinking%20-%20employee%20blog-gers.pdf

Edelman & Intelliseek. (2005). *Insights* "Trust MEdia" How real people are finally being heard. Retrieved from http://www.brickmeetsbyte.com/images/uploads/ISwp_TrustMEdia_FINAL.pdf

Foremski, T. (2006, Feb. 27). Die press release die! Retrieved from http://www.siliconvalleywatcher.com/mt/archives/2006/02/die_press_relea.php

Gogol, P. (2006, October 17). Wal-Mart vs. the blogosphere. *Businessweek.* Retrieved from http://www.businessweek.com/stories/2006-10-17/wal-mart-vs-dot-the-blogospherebusi-nessweek-business

Grier, C., Thomas, K., & Nicol, D. (2010, January). Barriers to security and privacy research in the web era. *imchris.org.* Retrieved from http://www.inwyrd.com/blog/wp-content/uploads/2010/03/wecsr-submit.pdf

Gross, G. (2009, October 5). FTC: Bloggers must disclose payments for endorsements. *PC World.* Retrieved fromhttp://www.pcworld.com/article/173127/article.html

IABC. (March, 2008). Social media releases for IABC. Retrieved from http://news.iabc.com/index.php?s=43&item=137

Key, R.J. (2005). How the PR profession can flourish in this new digital age: Why you must challenge old PR models, *Public Relations Tactics*, November, pp. 18–19.

Lattimore, D., Baskin, O., Heiman, S.T., &Toth, E.L. (2009). *Public relations: The profession and the practice*. New York: McGraw-Hill Higher Education.

Li, C., & Bernoff, J. (2008). *Groundswell, winning in a world transformed by social technologies*. Cambridge, MA: Forrester Research Inc.

Madden, M. (2009, July 29). The audience for online video-sharing sites. Pew Internet & American Life Project. Retrieved from http://pewinternet.org/Reports/2009/13—The-Audience-for-Online-VideoSharing-Sites-Shoots-Up.aspx

McCorkindale, T. (2010). Can you see the writing on my wall? A content analysis of the Fortune 50s Facebook social networking sites. *Public Relations Journal, Vol. 4, No. 3* Retrieved from http://www.prsa.org/intelligence/prjournal/documents/2010mccorkindale.pdf

Mendelsohn, J. (2010, March). Honey, don't bother mommy. I'm too busy building my brand. *New York Times*. Retrieved from http://www.nytimes.com/2010/03/14/fashion/14moms. html?pagewanted=all&_r=0

Mui, Y. Q., & Whoriskey, P. (2010, December 30). Facebook passes Google as most popular site on the Internet. *Washington Post*. Retrieved from http://www.washingtonpost.com/wp-dyn/content/article/2010/12/30/AR2010123004645.html

PRSA Code of Ethics. (2012). Retrieved from http://www.prsa.org/AboutPRSA/Ethics

Schneier, B. (2010, July/August). A taxonomy of social networking data. Retrieved from http://www.schneier.com/essay-322.html

Smith, A. (2011, January 27). 22% of online Americans used social networking or Twitter for politics in 2010 campaign. Pew Internet & American Life Project. Retrieved from http://www.pewinternet.org/reports/2011/politics-and-social-media.[

Solis, B., & Breakenridge, D. (2010). *Putting the public back in public relations: How social media is reinventing the aging business of PR*. Upper Saddle River, NJ: Pearson Education.

Stone, B., & Richtel, M. (2007, July 16). The hand that controls the sock puppet could get slapped. *New York Times*. Retrieved from http://www.nytimes.com/2007/07/16/technology/16blog.html?pagewanted=all

Swallow, E. (2010, July, 20). 10 tips for corporate blogging. *Mashable*. Retrieved from http://mashable.com/2010/07/20/corporate-blogging-tips/

Technorati's State of the Blogosphere. (2011). Retrieved from http://technorati.com/state-of-the-blogosphere

Weaver Lariscy, R., Avery, E. J., Sweetser, K.D., & Howes, P. (2009) Monitoring public opinion in Cyberspace: How corporate public relations is facing change. *Public Relations Journal, Vol 3. N0.4*, Retrieved from http://www.prsa.org/SearchResults/view/6D-030406/0/Monitoring_Public_Opinion_in_Cyberspace_How_Corpor

Wilcox, D. L., & Cameron, G. T. (2010). *Public relations strategies and tactics*. Boston, MA: Allyn & Bacon.

Wright, D., & Hinson, M. D. (2009). *An analysis of the increasing impact of social and other new media on public relations practice*. Retrieved from http://www.instituteforpr.org/topics/analysis-of-social-and-other-new-media/

Wright, D. K., & Hinson, M.D. (2008). How blogs and social media are changing public relations and the way it is practiced. *Public Relations Journal, 2(2)*.Retrieved from http://www.prsa.org/SearchResults/view/6D-020203/0/How_Blogs_and_S

# · 1 1 ·

# Toward a New
# Code of Ethics

## Social Media in Journalism

KELLY FINCHAM

Social media expert and Columbia professor Sree Sreenivasan has described the social media networks such as Twitter and Facebook as the "biggest advance in the Internet for journalists since the debut of the public web," (Sreenivasan, 2010). Sreenivasan says social media improves the journalist's toolbox by providing the following:

1. Access to new sources, new ideas, new topics, new trends.
2. Access to existing sources.
3. Access to new eyeballs and attention to your work.
4. Access to creating, enhancing, and curating your online brand.

Social media platforms such as Twitter and Facebook certainly offer unparalleled means of communicating with the public. Journalists use Facebook and Twitter to disseminate information, source stories and engage with their audience. Twitter has morphed into the "go-to source" for instant information even though it was originally designed as a short messaging service (O'Connor, 2009). The service allows users to send 140-character messages via cellphone or text and users can monitor feeds on their mobile device or computer.

Journalists now use it for newsgathering purposes such as searching for sources and story ideas and also to disseminate information. It's like the old AP wire feed except it's accessible to anyone with an Internet connection.

But these improved means of access have created new ethical issues for journalists, issues that would have been unthinkable just a few years ago. Part of the problem is that journalists are playing catch-up on Twitter and Facebook in the mistaken belief that they are behind the times. But as Sreenivasan says, social media in 2012 is where radio was in 1912 and TV in 1950. We will all be playing catch-up for some time.

Twitter, which was launched in July 2006, has evolved into a must-check destination for news and information. The service had been in existence for less than a year when the fatal campus shootings took place at Virginia Tech in April 2007. The Virginia Tech massacre was a watershed moment for social media as "mainstream American news channels woke up to the immediacy and power of personal accounts on Facebook, Flickr, Myspace and Twitter" (Jarvis & Barkham, 2007).

From the so-called Miracle on the Hudson in 2009 to the Japanese earthquake and tsunami and earthquake in 2011 to the tornados across the United States in 2011, Twitter has provided journalists and citizens with an unequalled ability to get the news instantly.

Indeed, *New York Times* reporter Brian Stelter used Twitter to report from Joplin, Missouri, in 2011, when he landed in the tornado-hit town without even a pen or a notebook. "Looking back, I think my best reporting was on Twitter," he said (Ingram, 2011). But Twitter's always-on platform, which facilitates the instant dissemination of verified and unverified information from a diverse range of sources (Hermida, 2010), is potentially problematic because of that instant dissemination.

For instance, serious errors were made on January 8, 2011, when it was reported on Twitter that U.S. Congresswoman Gabrielle Giffords had been shot dead during a constituency event in Arizona. Several news outlets including NPR, Reuters and CNN retweeted the erroneous report that Giffords was dead (Safran, 2011). Hermida (2011) says that the shooting showed how Twitter has transformed the process of journalism from "filter, then publish," to "publish, then filter."

But, just as Twitter facilitated the instant dissemination of erroneous information, it also allowed publications to immediately issue corrections. As Mallary Tenore (2011) put it, Twitter offers an improved real-time correction experience over the legacy print media, "which would have to wait until the next day's paper to publish a correction."

The Giffords case is a clear example of how this instant, real-time process of journalism has created new ethical dilemmas for journalists. And while social media may have advanced journalism, it has also changed the rules in reporting. Many journalists are being forced to play catch-up in the absence of any clear guidelines.

As the American Society of Newspaper Editors said in May 2011 platforms such as Facebook and Twitter offer "exciting opportunities" for reporters and news organizations but "they also carry challenges and risks."

Twitter is particularly challenging, as there "are still multiple questions around professional journalists' activities on Twitter that require thoughtful, open debate" (Posetti, 2009).

Posetti (2009) found that Twitter had become so entrenched in the reporting process that she advised media organizations to update their "existing editorial guidelines to make them relevant to social media platforms like Twitter." Three years on, Posetti's advice has been largely ignored. There is no consensus about social media ethics within the organizations that represent journalists or the organizations that produce journalism and many still cling to the "one-size-fits-all" approach of legacy ethics codes (Ward, 2011) instead of a more comprehensive framework.

In a 2011 review of American newspapers, the American Society of Newspaper Editors (ASNE) warned journalists about the two-edged sword of social media. "Putting in place overly draconian rules discourages creativity and innovation, but allowing an uncontrolled free-for all opens the floodgates to problems and leaves news organizations responsible for irresponsible employees" (ASNE, 2011). However, there is little guidance for journalists who are caught in this delicate balancing act. The ASNE issued its own best practice frameworks in May 2011, but its review was restricted to newspapers. This study extends that review process to national broadcasters and national journalism organizations in the United States (U.S.). For reasons of comparison, we included the relevant counterparts in the United Kingdom (U.K.).

Thus, this study offers a first step in creating that framework with a review of publicly available guidelines from the mainstream media and professional organizations in the U.S. We compare and contrast the guidelines in an attempt to identify the best practices in social media ethics.

To specify, we reviewed the following 14 media organizations:

**Newspapers**
*The Washington Post*
*The New York Times*

*The Wall Street Journal*
*USA Today*
*The Guardian* (U.K.)

**Broadcasters**
Fox News
CBS News
CNN
ABC News
NBC News
NPR
BBC (U.K.)
Sky News (U.K.)
We extended that review to the following four professional organizations:

**Professional Organizations**
The Radio Television Digital News Association (RTDNA)
The Online News Association (ONA)
The Society for Professional Journalists (SPJ)
The National Union of Journalists (NUJ/U.K.)

Taking the newspapers first, *The Washington Post*, *The New York Times* and *The Guardian* make their policies publicly available online, while *The Wall Street Journal* and *USA Today* don't. But a closer look reveals that only *The Washington Post* includes specific language covering social media. *The Guardian's* policies are generalized and do not refer to Twitter or Facebook, while the guidelines at *The New York Times* were last updated in 2005, before the advent of Twitter or Facebook as journalism tools.

Further investigation reveals that the social media policies at *The Guardian*, *The Wall Street Journal* and *USA Today* are for internal use only. Thus, only one of the newspapers studied, *The Washington Post*, makes its clear and transparent guidelines publicly available online. This pattern is repeated across the broadcasting platforms. Of the eight broadcasters surveyed, only three posted public links to their social media policies. The three broadcasters are the BBC, NPR and NBC News. Of the remaining five broadcasters, the guidelines for CBS News, ABC News and Sky News (U.K.) are unavailable publicly; CNN's policies are available only on a fired employee's blog from 2008, and Fox says it does not give its staff guidelines.

In early 2012, Sky News, which is part of News Corporation, was singled out by a *Guardian* reporter for its "draconian" views on social media after an internal e-mail revealed that the broadcaster had banned staffers from breaking news on Twitter or retweeting anyone who did not work for Sky (Holliday, 2012. The guidelines also told reporters to "stick to their beat" and to refrain from tweeting about personal issues on their professional accounts. De Rosa (2012) says such policies would drive Sky into obsolescence.

Sky might do well to learn from the BBC, where social media editor Chris Hamilton (2012) summarizes the organization's approach as "don't do anything stupid." And Sky's approach is in sharp contrast to NPR, which advises reporters to treat social media like a gracious host (Sonderman, 2012).

This differing treatment of social media culture is at the heart of an increasingly fractured debate over social media policies. While organizations such as the U.K.'s Sky News treat social media with a "Victorian internet" mentality (De Rosa, 2012), others, notably the BBC, NPR and *The Washington Post*, embrace the platform.

In the U.S., NPR has been hailed for its "refreshing" social media policy (Sonderman, 2012), which it released in February 2012 to specifically include platforms such as Facebook and Twitter. Sonderman (2012), the Digital Media Fellow at the Poynter Institute, says other organizations should treat the NPR code as a social media "blueprint," in particular for the way it advises staff to respect the culture of the Internet.

This difference in approaches raises the question of whether the standard rules of ethics can serve the new media for journalistic purposes or whether journalistic ethics have to evolve because of the way social media is transforming journalism. The standard journalism ethics are perhaps best explained by the code from the Society for Professional Journalists (SPJ), which says that members should

> Seek truth and report it
> Minimize harm
> Act independently
> Be accountable

Under "seek truth and report it," the SPJ guidelines offer very specific advice in areas involving print and broadcast media. For instance, it cautions that journalists should not misrepresent through the use of "headlines, news teases and promotional material, photos, video, audio, graphics, sound bites and

# The 10 Commandments

1. Don't do anything stupid.
2. Do identify yourself as a journalist.
3. Do assume that everything you post is public, regardless of privacy controls.
4. Don't share or retweet anything that creates questions about your impartiality.
5. Don't share or retweet anything that creates questions about conflict of interest.
6. Do fact-check before you share information.
7. Do be bi-partisan in "Following" and "Friending."
8. Do issue corrections quickly.
9. Don't delete a tweet, even if it's wrong.
10. Do be careful about copying information from social media profiles.

## 1. Don't do anything stupid

"Don't do anything stupid," is a universal theme across the five organizations.

- The BBC uses "Don't do anything stupid," as its jumping-off point, saying that it summarizes the overall policy. It says that even when reporters are using social media for personal use, they should remember that their personal groups of friends and contacts still view them as a BBC representative.
- NBC News says that while it "encourages participation in social media," it requires employees to recognize that they are representing the broadcaster at all times.
- *The Washington Post* advises journalists that their social media accounts, whether personal or work-related, "reflect upon the reputation and credibility of *The Washington Post's* newsroom" and that journalists must always protect their professional integrity.
- RTDNA says journalists should never post anything that might embarrass them and always to remember that, "when you work for a journalism organization, you represent that organization on and off the clock."
- NPR advises journalists to "tread carefully "and proceed with caution.

All the organizations listed advise reporters not to share or retweet any private information about their employer. This advice stems from an incident in May

2009, at *The New York Times* when an internal meeting was tweeted by, among others, then staff-reporters Jennifer 8. Lee and Brian Stelter (*New York Observer*, 2009).

This is the headline advice from all four news organizations:

- "You are on show to your friends and anyone else who sees what you write, as a representative of the BBC.
- "And always remember, you represent NPR."
- "You are a representative of NBC News."
- "*Washington Post* journalists are always *Washington Post* journalists."
- RTDNA: "You represent your organization on and off the clock."

Possibly the best explanation for the "don't do anything stupid" rule comes from NPR, which says the following:

> The Internet and the social media communities it encompasses can be incredible resources. They offer both a remarkably robust amount of historical material and an incredible amount of "real-time" reporting from people at the scenes of breaking news events. But they also present new and unfamiliar challenges, and they tend to amplify the effects of any ethical misjudgments you might make. So tread carefully. Conduct yourself online just as you would in any other public circumstances as an NPR journalist. Treat those you encounter online with fairness, honesty and respect, just as you would offline. Verify information before passing it along. Be honest about your intent when reporting. Avoid actions that might discredit your professional impartiality. And always remember, you represent NPR.

The evolution of the "always-on-journalist" has been accelerated by the access that social media provides. The "people formerly known as the audience" (Rosen, 2006) are now engaged with reporters in ways that would have been unimaginable even in 2006. A reporter is visible across social media platforms and it's a two-way street. The journalist can see the audience, and the audience can see the journalist. As the RTDNA describes it, social media "narrows the distance between journalists and the public."

In addition, journalists' professional and personal lives are merging in ways that would be unrecognizable to journalists from as recently as the 1990s. RTDNA even advises that newsroom employees recognize that their words, although nominally private, are "direct extensions of their news organizations." And RTDNA points out that new technological developments mean that words can literally be tagged and traced to their source.

## 2. Do identify yourself as a journalist

The standard advice here is to never use an anonymous identity on any social media site, whether in the newsroom or on a personal site, and to always specify your work affiliation.

- RTDNA cautions that journalists must, and will, be held responsible for everything they say. It advises that journalists should never use anonymous user names or avatars. "You are responsible for everything you say. Commenting or blogging anonymously compromises this core principle."
- NBC News warns reporters to use their official e-mail and always disclose who they are working for. It also posts very specific guidelines for reporters who may be looking to garner information anonymously and tells reporters that they must seek approval for such activities from the Executive Producer for Standards or the Editor in Chief.
- NPR requires staff to identify themselves as NPR journalists on all social media platforms when working and bans them from using pseudonyms. There is a gray area at NPR around the use of anonymous names on certain sites. NPR allows for anonymous identities as long as any information gleaned during such interactions is not used in any subsequent reporting. However, this approach could potentially be problematic and should be avoided.
- *The Washington Post* does not address the anonymity issue but does advise journalists to ensure that their social media bios reflect their occupation. "Post journalists are free to continue using personal accounts but should remember that they remain, at all times, Washington Post journalists."
- The BBC, alone of the organizations surveyed, does not issue guidelines for personal accounts as long as the account holder is not identified as a BBC employee. In other words, the guidelines do not apply to BBC employees who do not identify themselves as such, "do not discuss the BBC and are purely about personal matters."

However, this may be a matter of shading, as the summary guidelines tell journalists that, "if you are editorial staff it doesn't make much difference whether or not you identify yourself as someone who works for the BBC." In other words, if you work in the editorial department at the BBC, you should always identify yourself as a journalist.

## 3. Do assume that everything you post is public, regardless of privacy controls

Raju Narisetti, now the Managing Editor for Digital at *The Wall Street Journal*, is a high-profile example of how private accounts fail to protect journalists from public scrutiny. In 2009, while working as one of two managing editors at *The Washington Post*, he posted several "private" tweets from his then-protected Twitter account, including these two: "We can incur all sorts of federal deficits for wars and what not. But we have to promise not to increase it by $1 for healthcare reform? Sad." and "Sen Byrd (91) in hospital after he falls from 'standing up too quickly.' How about term limits. Or retirement age. Or commonsense to prevail" (Hohmann, 2011).

Narisetti's comments were distributed far beyond his 91 Twitter followers on his private account, which he closed as a result of the controversy they created. The Post Ombudsman wrote the following:

> Narisetti's tweets could be seen as one of The Post's top editors taking sides on the question of whether a health-care reform plan must be budget neutral. On Byrd, his comments could be construed as favoring term limits or mandatory retirement for aging lawmakers. Many readers already view The Post with suspicion and believe that the personal views of its reporters and editors influence the coverage. The tweets could provide ammunition. (Alexander, 2009)

That episode, which prompted *The Washington Post* to expedite the release of its social media guidelines, goes to the heart of the privacy problem. Effectively, there is no private space online for employees at news organizations. This points to the conflict between the protection of free speech and the ethical issues for journalists on social media. As Narisetti tweeted before he closed his account. "For flagbearers of free speech, some newsroom execs have the weirdest double standards when it comes to censoring personal views" (Alexander, 2009).

*The Washington Post* issued its revised guidelines shortly after the Narisetti affair. The guidelines are practical in approach and the paper acknowledges that the guidelines are evolving along with the platform and will be revised accordingly. But they emphasize that "every comment or link we share should be considered public information regardless of privacy settings."

It's the same story at NPR, which again states that all content on a social media website should be considered public. "Anyone with access to the web can potentially see what we're doing." As NPR puts it, "Regardless of how careful we are in trying to keep them separate, our professional lives and our personal lives overlap when we're online."

NPR advises staff to use the highest level of available privacy tools but warns that the line between public and private has become so blurred by social media that even personal messages to friends and family can be "easily circulated beyond the intended audiences."

NPR uses a fictitious case study to illustrate how good tweets can go bad. It describes how NPR legal correspondent and college basketball fan Sue Zemencourt, gets involved in an online row over college basketball. Zemencourt's comments are heated and intemperate, so much so that the blog blocks any further comments from her and also discovers that her IP address leads back to NPR. The blog's host posts a comment along the lines of, "someone at NPR is using language that the FCC would not approve of." In the fictitious example, the controversy goes viral and *The Daily Show* weighs in. This simple illustration is a useful way of explaining to reporters why private tweets can cause public scandals.

The RTDNA guidelines reiterate this. They warn journalists that everything posted online is "open to the public—even if you consider it to be private." And the RTDNA goes further than the news organizations in warning journalists that, "Newsroom employees should recognize that even though their comments may seem to be in their 'private space,' their words can become direct extensions of their news organizations."

The RTDNA guidelines warn reporters that basic technology tools such as search engines and social mapping can easily link posts—even anonymous posts—to people, and thus their employers.

The technological reasons for the lack of privacy on the web stem from the guidelines that state, "Search engines and social mapping sites can locate their posts and link the writers' names to their employers." It extends this advice even when the reporter does not name names. It uses the example of a fictitious reporter who posts on his Facebook page that he is covering a fictitious mayoral candidate who is "dumber than dirt." The candidate's campaign staff demand that the reporter "be taken off the campaign," while the reporter's defense is that his words were private and he didn't name names.

NBC News agrees that this is an issue. "Be aware that it is virtually impossible to post items anonymously," it says in the guidelines. "There have been numerous examples of inappropriate postings that were intended to be anonymous, but were traced back to the individual doing the posting or to the company computer from which the post was made." And it warns staff that, "the line between the public and private is increasingly blurred. You would be wise to presume that nothing on the Internet is either private or confidential."

## 4. Don't share or retweet anything that creates questions about your impartiality

News organizations are increasingly uncomfortable with editorial staff taking any kind of position on public issues. Caitlin E. Curran, a freelance producer for NPR, found this out the hard way when she was fired from her part-time position at The Takeaway after a single photograph of her at an Occupy Wall Street rally appeared in the media in October 2011.

The photograph, which was taken by photographer Ben Furnas at a protest in New York City on October 15, 2011, "went viral" on Twitter (Joyner, 2011). Jennifer Houlihan, a spokeswoman for WNYC—which co-produces The Takeaway with Public Radio International—said Curran was in clear breach of the editorial guidelines: "When Ms. Curran made the decision to participate in the protest and make herself part of the story, she violated our editorial standards" (Estes & Grandoni, 2011).

Curran protested that she was attending the rally as an observer, but the photograph ultimately proved too damaging. This example underscores the issues surrounding the appearance of impartiality for journalists when related to political and controversial issues.

In Britain, the BBC cautions staff that their personal social media profiles are considered "representative of the BBC" and that they should temper their comments accordingly. It advises staff to maintain impartiality by not revealing their political preferences. At NBC News, the guidelines specifically prohibit staff from taking public positions on any "controversial or political issues" unless such positions have been approved in advance.

The RTDNA says reporters must be careful when writing, tweeting or blogging about news: "Editorializing about a topic or person can reveal your personal feelings." And it warns that some kinds of biased comments could land the journalist in court on libel charges. *The Washington Post* says journalists must not tweet or posts words, video or images that "could be perceived as reflecting political bias or favoritism."

"When posting content online, ask yourself: Would this posting make a reader question my ability to do my job objectively and professionally (whether you are a reporter, an editor, a developer or a producer)? If so, don't post it."

## 5. Don't share or retweet anything that creates questions about conflict of interest

The standard advice is always to disclose any personal or business relationships, and RTDNA is particularly useful on this issue, as it warns reporters that they can "not use social media to promote business or personal interests without disclosing that relationship to the public."

Social media has amplified the potential conflicts of interest. Consider this example from RTDNA: How does a news organization acknowledge the business relationship with an advertiser who has also been the subject of a glowing review for one of its new products?

This applies to much more than just the obvious business relationships. The BBC uses the example of staffers being asked to produce any content for payment for blogs or micro-blogs. It says any such requests must be cleared with each employee's line manager.

*The Washington Post* requires that journalists "not accept or place tokens, badges or virtual gifts from political or partisan causes on pages or sites, and should monitor information posted on their public profiles (by individuals or organizations) for appropriateness." This area is best addressed by the RTDNA.

## 6. Do fact-check before you share information

NPR says that when its journalists post information being reported on social media sites, they are "effectively reporting that information" themselves. It says the best policy in this instance is to be transparent. "Tell readers what has and hasn't been confirmed." It asks journalists to consider if they are about to "spread a thinly-sourced rumor or am I passing on valuable and credible (even if unverified) information?"

*The Washington Post* says that accuracy is the "over-riding concern" and that stories must be evaluated before being posted. If the story cannot be evaluated, then editors must look at the sourcing. If the source is reliable, the story can be cited and attributed to the source. If the facts or sourcing are what the *Post* terms *murky,* the guidelines advise that reporters "buy time by telling readers we're investigating a developing story."

Many of these guidelines were developed after the fatal shootings at a community event hosted by Congresswoman Gabrielle Giffords (D-AZ) in January 2011. Several news organizations tweeted the mistaken report that Giffords was dead, including NPR and *The Washington Post.*

While it was difficult for the national outlets to fact-check the story as it was breaking, local news outlets were not making the same kind of reporting errors (Tenore, 2011). In its guidelines, *The Washington Post* now says that it should have prefaced its tweet reports with "CNN and NPR are reporting that Rep. Gabrielle Giffords has died after being shot in the head by a gunman."

Their tweet at the time read as follows: "Rep Gabrielle Giffords has died after being shot in the head by a gunman, according to NPR and CNN."

The RTDNA takes the view that Twitter and Facebook need to be treated like more traditional crowd-sourcing techniques like police scanners or phone tips. "If you cannot independently confirm critical information, reveal your sources: tell the public how you know what you know and what you cannot confirm," it says. It also says that Twitter's "character limits and immediacy are not excuses for inaccuracy and unfairness."

## 7. Do be bi-partisan in "Following" and "Friending"

The term "friend" should probably be changed to "connections" on Facebook. Rather than "friending" people (and creating the appearance of the type of relationship the word implies), the activity would be better described as "connecting." Until then, journalists should tread warily.

As RTDNA says, "You may believe that 'online' friends are different from other friends in your life, but the public may not always see it that way." The organization advises journalists to check their "friends" lists regularly to try and avoid any conflicts with people who "become newsmakers." It offers the example of a news editor who becomes friends with a neighbor on Facebook. One year later the neighbor runs for mayor and the current mayor is upset because his challenger shows up as a friend on the reporter's Facebook account. What should the reporter do?

NPR would recommend that the journalist "friend" the mayor as a way of ensuring impartiality on the issue. NPR takes the view that the use of Facebook in this way is "as basic a tool as signing up to be on mailing lists used to be." NPR also makes it quite clear that journalists may follow or friend political parties or advocacy groups but that they can only do so to be an observer, not a participant.

*The Washington Post* requires reporters to seek specific permission if they "join, follow or friend any person or organization online." This is a sensitive area for all news organizations. The BBC says, "It may be appropriate to join Facebook groups related to political causes for reasons of political research. Where this is agreed we should be transparent and should consider how membership of the group can be balanced." It cites the example of a reporter joining the Facebook groups for the four main British political parties: the Conservatives, Labour, Liberal and the Nationalists. But the issue is serious enough for the BBC to require that reporters "in politically sensitive areas" check in advance with their managers.

The NPR advice is probably the clearest, and most sensible, in this example. It advises that reporters "follow" or "friend" political parties and advocacy groups as long as they are doing it to "keep up on what that party or group is doing." And, NPR says, reporters "should be following those on the other side of the issues as well."

## 8. Do issue corrections quickly

*The Washington Post* advises reporters to "take ownership" of the situation. "If you mistakenly retweet or forward erroneous information, correct your mistake in a subsequent tweet/update and make an effort to provide a more accurate link." It's the same advice at RTDNA, which reminds reporters that social media is also a vast online archive where mistakes live on forever. "Correct and clarify mistakes, whether they are factual mistakes or mistakes of omission."

A clear example of this can be seen in the hours after the Giffords shooting, as news outlets began to realize the congresswoman had survived the attack. Andy Carvin at NPR, who had originally tweeted that Giffords had died, began sending out amended tweets. He later explained his actions in this way:

> With around 2 million people following @nprnews and @nprpolitics, deleting the original tweet wouldn't have altered the fact that many of those followers had already seen the mistaken tweet and retweeted it. So based on that reasoning, I (Carvin) decided to be transparent about the mistake and not try to hide. Was that a good or a bad decision? At the time I felt it was a reasonable decision, given the circumstances, and still feel comfortable with the decision. I can imagine if I had deleted it, we'd be reading news stories and blog posts today about NPR trying to cover our tracks on Twitter. (Safran, 2011)

The BBC was among the organizations that reported the incorrect information about Giffords, and it too decided to post updated tweets without pulling the earlier ones. The BBC's Alex Gubay agrees with Carvin:

> As many other news organizations did, we believed the information to be correct at the time, but updated accordingly as soon as we realized that was not the case. In terms of deleting tweets, we didn't feel it appropriate or helpful to do so, and that remains the case now. (Safran, 2011)

## 9. Don't delete a tweet, even if it's wrong

NPR's Carvin is considered to be an authority on social media. Speaking in 2011, he stood behind NPR's decision not to delete the original tweet reporting that Giffords had been killed:

I posted that she had been killed because that is what we were reporting, and as soon as I saw we were backing off from that assertion, I posted the followup noting that as well. I very briefly considered deleting the incorrect tweet, but concluded it was both pointless and inappropriate. (Safran, 2011)

However, three organizations, PBSNewsHour, CNN and Reuters, which tweeted the wrong information about Giffords, subsequently deleted the tweets. Teresa Gorman at PBS Newshour explained the decision on Twitter that "in all transparency, we did tweet NPR's news, but I chose to delete it because it kept getting retweeted hours later" (Silverman, 2011).

WBUR in Boston is part of the NPR family and it too retweeted that Giffords was dead. Speaking in 2011, WBUR's Andrew Phelps said:

We have decided NOT to delete the erroneous tweet, because it serves as part of the narrative of this story. Facts can change fast when news is breaking, and that leads to errors. We need to own the error, not hide from it. But we also need to rectify the error and explain ourselves to people who trust us. Deleting the tweet would do more to harm trust than preserving it would do to harm truth.

Perhaps this line, more than any, should guide editorial decisions about deleting erroneous tweets: "Deleting the tweet would do more to harm trust than preserving it would do to harm truth" (Phelps, 2011).

## 10. Do be careful about copying information from social media profiles

Many newsrooms and journalists grapple with the question of whether to use pictures and information from a private individual's account if and when that person becomes part of a news story. For example, reporters on the scene of a campus shooting may try to "friend" students in an attempt to capture more material about the event or any victims.

The consensus from this study is that reporters should only use such information when there is a reasonable expectation that the person concerned believed that this information would be made public.

Again, the RTDNA offers the clearest direction on this issue as the news organizations are primarily concerned with what their reporters are posting on the social networks. The RTDNA offers this checklist for reporters, which is based on its advice for journalists who are working undercover:

~ Does the poster have a 'reasonable expectation' of privacy?
~ Is this a story of great significance?

~   Is there any other way to get the information?
~   Are you willing to disclose your methods and reasoning?
~   What are your journalistic motivations?

## Conclusions

In the absence of a clearly defined set of external guidelines for the professional organizations, internal rules are evolving at different news organizations, which are creating tensions for journalists. This study shows that professional organizations such as the SPJ must take the lead in this discussion and should clarify and codify new rules that will help journalists maintain ethical standards in social media.

The study also points to the necessity for news organizations to create specific posts for social media editors who manage the newsrooms' social media.

### References

Alexander, A. (2009). Post editor ends tweets as new guidelines are issued. *Washington Post.* Retrieved from http://voices.washingtonpost.com/ombudsman-blog/2009/09/post_editor_ends_tweets_as_new.html

Bauder, D. (2010). Octavia Nasr fired by CNN—the editor tweeted admiration for Grand Ayatollah Mohammed Hussein Fadlallah. *Washington Post.* Retrieved from http://www.washingtonpost.com/wp-dyn/content/article/2010/07/07/AR2010070704948.html

Buttry, S. (2010, April 21). Journalists' code of ethics: Time for an update? http://stevebuttry.wordpress.com/2010/11/07/journalists-code-of-ethics-time-for-an-update/

De Rosa, A. (2012). Sky News longs for Victorian internet, applies dark age social policy. *Reuters.* Retrieved from http://blogs.reuters.com/anthony-derosa/2012/02/07/sky-news-longs-for-victorian-internet-applies-dark-age-social-policy/

Estes, A. C., & Grandoni, D. (2011). Another public radio freelancer gets the ax over Occupy Wall Street. *Atlantic Wire.* Retrieved from http://www.theatlanticwire.com/national/2011/10/another-public-radio-freelancer-gets-ax-over-occupy-wall-street/44278/

Halliday, J. (2012). Sky News clamps down on Twitter use. *Guardian.* Retrieved from http://www.guardian.co.uk/media/2012/feb/07/sky-news-twitter-clampdown

Hamilton, C. (2012). Breaking news guidance for BBC journalists. *BBC Editor's Blog.* Retrieved from http://www.bbc.co.uk/blogs/theeditors/2012/02/twitter_guidelines_for_bbc_jou.html

Hermida, A. (2010). Twittering the news. *Journalism Practice, 4*(3), 297–308. doi:10.1080/17512781003640703

Hohmann, J. (2011). Best practices for social media. *ASNE.* Retrieved from http://asne.org/portals/0/publications/public/10_best_practices_for_social_media.pdf

Ingram, M. (2011). NYT reporter shows the power of Twitter. *Giga Om.* Retrieved from http://Gigaom.Com/2011/05/27/Nyt-Reporter-Shows-The-Power-Of-Twitter-As-Journalism/

Jarvis, J., & Barkham, P. (2007). Were reporters right to solicit information from students' web pages? *Guardian*. Retrieved from http://www.guardian.co.uk/media/2007/apr/23/monday-mediasection

Joyner, J. (2011). Signs your sign will go viral. *Outside the Beltway*. Retrieved from http://www.outsidethebeltway.com/signs-your-sign-will-go-viral/

Kanigel, R. (2012). *The Student Newspaper Survival Guide.*, Malden, MA: Wiley Blackwell, p. 258.

*New York Observer*. (2009). Twitter culture wars at The Times: 'We need a zone of trust,' Bill Keller tells staff. Retrieved fromhttp://www.observer.com/2009/05/twitter-culture-wars-at-ithe-timesi-we-need-a-zone-of-trust-bill-keller-tells-staff/

O'Connor, R. (2009). Twitter journalism. Huffington Post. Retrieved from http://www.huffingtonpost.com/rory-oconnor/twitter-journalism_b_159101.html

Phelps, A. (2011). Why we didn't delete the Tweet. *Hubbub*. Retrieved from http://hubbub.wbur.org/2011/01/10/giffords-coverage-twitter

Posetti, J. (2009). Rules of engagement for journalists on Twitter. *PBS Media Shift*. Retrieved from http://www.pbs.org/mediashift/2009/06/rules-of-engagement-for-journalists-on-twitter170.html

Rosen, J. (2006). The people formerly known as the audience. *PressThink*. Retrieved from http://archive.pressthink.org/2006/06/27/ppl_frmr.html

Safran, S. (2011). How an incorrect report of Giffords' death spread on Twitter. *Lost Remote*. http://www.lostremote.com/2011/01/09/how-an-incorrect-report-of-giffords-death-spread-on-twitter/#comment-126540166

Silverman, C. (2011). To delete or not to delete? *Columbia Journalism Review*. Retrieved from http://www.cjr.org/behind_the_news/to_delete_or_not_to_delete.php

Sonderman, J. (2012). NPR's new guidelines for using social networks: 'Respect their cultures.' *Poynter*. Retrieved from http://www.poynter.org/latest-news/media-lab/social-media/164202/npr-new-guidelines-for-using-social-networks-respect-their-cultures/

Sreenivasan, S. (2010, July). *Advanced social media for journalists*. Paper presented at the Washington Foreign Press Corps, Department of State, Washington, DC. Retrieved from http://fpc.state.gov/139966.htm

Tenore, M. (2011). Conflicting reports of Giffords' death were understandable but not excusable. *Poynter*. Retrieved from http://www.poynter.org/latest-news/top-stories/113876/conflicting-reports-of-giffords-death-were-understandable-but-not-excusable/

Ward, S. (2011). Rethinking journalism ethics, objectivity in the age of social media. *PBS Media Shift*. Retrieved from http://www.pbs.org/mediashift/2011/07/rethinking-journalism-ethics-objectivity-in-the-age-of-social-media208.html

## Social Media Guidelines

*The Washington Post* rules
    http://www.washingtonpost.com/wp-srv/guidelines/social-media.html
*The New York Times* rules
    http://www.nytco.com/press/ethics.html#B5

*The Guardian* rules

    http://www.guardian.co.uk/info/2010/oct/19/journalist-blogging-commenting-guidelines

NBC News rules

    http://msnbcmedia.msn.com/i/MSNBC/Sections/AboutUS/NBC/NBCNews_online_

    publishing_guidelines.pdf

BBC rules

    http://www.bbc.co.uk/editorialguidelines/page/guidance-blogs-bbc-summary

NPR rules

    http://ethics.npr.org/tag/social-media/

RTDNA rules

    http://www.rtdna.org/pages/media_items/social-media-and-blogging-guidelines1915.php

American Society of Newspaper Editors. Best practices for social media. *ASNE*. Retrieved from

    http://asne.org/portals/0/publications/public/10_best_practices_for_social_media.pdf

Society of Professional Journalists, Code of ethics, http://www.spj.org/ethicscode.asp

# · 1 2 ·

# Epilogue

## Back to the Future

SUSAN J. DRUCKER & GARY GUMPERT

An epilogue is more than an afterthought. Like the coda, its musical equiva‑
lent, it ought to bring to a close the intertwining themes that preceded it. If
there is a reverberating motif to this book about the regulation of social media,
it has something to do with the accelerating impact of technological change
upon social relationships and the ethics and regulations that accompany
change and function. One would assume that this extraordinary rate of inno‑
vation would have reached a degree of deceleration, but there appears to be no
end in sight. The impact of communication technology on our daily existence
is simply mind-boggling. Who would have thought 25 years ago that the tele‑
phone (if one can call the modern instrument that) would be used to navigate
our movement? Who would have imagined that the same phone would be used
as a camera? Who would have imagined that global connection would be an
expectation rather than a mystical dream? The fusing and miniaturization of
the computer, telephone, radio, television continue to evolve. The need to
incorporate such innovation into our daily lives is and must be accompanied
by a regulatory system that encourages further innovation and serves to pro‑
tect both citizen communication and industry.

The preceding pages have provided a snapshot, framed by the constraints
and realities of publishing, because by the time this volume is actually published,

new innovations and further regulatory issues will have surfaced. A still life of a moment in motion is both limited and illuminating. The epilogue allows for the careful consideration of what came before and prompts prediction of what will come next. Attempting to forecast the future is, by definition, an ultra hazardous activity yet, with a mixture of humility and hubris, we feel compelled to do just that based on the intertwined themes and threads emerging from the pages of this volume. What follows are three observations about the future.

## Observation 1: The concept of medium will recede into the background as content takes center stage in the foreground.

A medium is by definition technologically based. The media—telegraph, radio, telephone, television, facsimile, computer, iPhone, iPod, et al.—are technological marvels and in their process of fusing and converging, become even more technically intricate. Paradoxically, their unique intricate structure begins to recede into the background as they advance along a continuum of intricate virtuosity. On the surface they are less visible, while the access to content of information and data becomes increasingly prominent. The ultimate user interface of technology and data suggests the disappearance of device. Taken to an extreme, predictions of implants replacing external devices abound. Ultimately, will Facebook morph into facechip (H. Cohen, personal communication, February 12, 2012)?

Back in 1996, Bill Gates wrote "Content Is King." In this article, he said the following:

> Content is where I expect much of the real money will be made on the Internet, just as it was in broadcasting. The television revolution that began half a century ago spawned a number of industries, including the manufacturing of TV sets, but the long-term winners were those who used the medium to deliver information and entertainment. When it comes to an interactive network such as the Internet, the definition of "content" becomes very wide. For example, computer software is a form of content—an extremely important one, and the one that for Microsoft will remain by far the most important. But the broad opportunities for most companies involve supplying information or entertainment. No company is too small to participate. One of the exciting things about the Internet is that anyone with a PC and a modem can publish whatever content they can create. In a sense, the Internet is the multimedia equivalent of the photocopier. It allows material to be duplicated at low cost, no matter the size. It will be about who owns content rather than who owns the medium.

If a medium recedes into the background, some governments are poised to regulate on the basis of content. Such regulation is antithetical to the current U.S. constitutional yardsticks used in determining the regulation of communication. The distinction between "content-based" and "content-neutral" categories is central to contemporary communication law. Erwin Chemerinsky (2000), a prominent scholar in U.S. Constitutional law, argues, "Today, virtually every free speech case turns on the application of the distinction between content-based and content-neutral laws."[1] Content neutral restrictions are also known as "time, place, and manner" regulation, thus stressing "medium" or "device" rather than "content."

The content-neutrality standard has become the touchstone of inquiry in free expression cases. A content-based regulation is one based on the substance of the message being communicated, rather than just the manner or method in which the message is being expressed. With a content-neutral restriction, the manner in which an expression is communicated or conveyed is central.

Key to this approach is the determination of categories of communication that have been deemed to lack social value or be undeserving of constitutional protection.[2] The strict scrutiny standard is applied to any government legislation that limits the content of protected speech (*Simon & Schuster, Inc. v. New York*, 1991). This means that in order for a regulation to survive strict scrutiny, "the State must show that its regulation is necessary to serve a compelling state interest and is narrowly drawn to achieve that end" (*Arkansas Writers' Project, Inc. v. Ragland*, 1987). The bottom line is that content-neutrality is king in U.S. law.

## Observation 2: The individual will function in multiple simultaneously co-existing communities.

Howard Cohen notes, "We are engaged in the culture of the self, now everybody can share with everybody or with as many 'anybodies' as they want" (H. Cohen, personal communication, February 12, 2012). Culture and the sharing of self is achieved through the fusing of multiple devices and the creation of placeless communities or communities based upon access rather nearness.

Traditional communities, those requiring some degree of physical contact among their constituents, also carried with it an obligation to "proximate others" in exchange for the benefit of membership. It should be noted that the term "contemporary community standards" became the standard legal test for obscenity first adopted in 1957 in *Roth v. U.S.* (1957) and further clarified in *Miller v. California* (1973). Generally, juries consist of and represent the community.

The contemporary community is complex as communication technologies facilitate the transcendence of place. Today's community of place is augmented and accompanied by aspatial communities electronically connected. The combination and fusing of place and worldwide connection has been referred to as "glocal" (local and global). Much social networking is designed to facilitate aspatial "egocentric" communities defined by function and interest. Therefore, the legal standard of community faces redefinition in an age of social media when the physical community of places exists along with virtual communities devoid of propinquitous obligation.

## Observation 3: A new government-industry regulatory paradigm will emerge.

As communication technologies are no longer restricted by a sense of place, when the regulation on the basis of location becomes an antiquated concept, the new regulation will constitute the fusion of technologies; the emergence of a supra national regulatory scheme based upon the interdependence of governments and global corporations will become preeminent.

Sergey Brin, the co-founder of Google, identified the threat to the freedom of the internet as coming from "a combination of governments increasingly trying to control access and communication by their citizens, the entertainment industry's attempts to crack down on piracy, and companies like Facebook and Apple, which tightly control what software can be released on their platforms" (Katz, 2012).

There are two distinct types of government regulatory threats emerging: the first refers to authoritarian regimes that ban access or threaten to do so, if foreign social media companies fail to comply with demands to filter, censor, or provide information on users; the second refers to the efforts of democratic governments who, in their support of communicative rights, threaten penalty and/or regulation.

We are familiar with the first type of direct regulation in which countries block or pressure media companies. According to the Open Net Initiative, many governments have filtered YouTube and others have blocked specific videos when the content violates national law. Copyrights, morality, and national security have all been used as justification regulation: filtering, banning, blocking.[3]

For example, in the United States, Senator Dick Durbin has suggested there is the "explicit threat that if Global Network Initiative (GNI) doesn't succeed in self-regulating the relationship between technology companies and repressive governments, Washington stands ready to regulate" (Downes, 2011). Resulting legislation could prohibit U.S. companies from doing business where fundamental U.S. values (including those in the First Amendment—freedom of the press—and Fourth Amendment—the right to be free from search and seizure) are violated. Both U.S. and European Union authorities have scrutinized behavioral advertising and its relationship to social media and social networking. While the United States and the European Union have typically taken very different approaches to privacy, with the United States following a market-driven approach to data privacy and the European Union a more direct approach, with laws emerging from European Union directives. A new European Union ePrivacy Directive requires consent for all cookies and other tracking codes, except those required to fulfill a direct user request (How to Comply, 2012).

In 2012, European Union members began investigating Google's new privacy policy. "France's data protection watchdog has cast doubt on the legality and fairness of Google's new privacy policy, which it said breached European law" (Pfanner, 2012). All this with the larger regulatory context in which the European Commission moved to overhaul its 17-year-old data protection rules in an effort to create more rigorous requirements.

> Under the proposed new EU rules, Internet companies like Google, Facebook and Yahoo would have to ask users whether they can store and sell their data to other businesses, such as advertisers, which is the source of almost all their income. Internet users can also ask for their data to be deleted from websites for good, the so-called "right to be forgotten." (E.U. Proposes, 2012)

In the face of emerging threats to the global reach of social media, the Global Network Initiative was launched in 2008. The larger context was one in which human rights groups and politicians condemned the top internet companies for complying with China's restrictive laws as a price of doing business in this massive market. Google had introduced a censored search engine in China (although the company has since shut down that site). Yahoo! had turned over data endangering Chinese activists. So the big three at the time (Google, Microsoft, and Yahoo!) established the Global Network Initiative in the oft-repeated pattern of self-regulation in the face of potential government regulation. "The initiative is modeled on previous voluntary efforts aimed at eradicating sweatshops in the apparel industry and stopping corruption in the oil, natural gas and mining industries" (Kopytoff, 2011).

The code proposed is a general code of conduct:

> The overarching commitment of the Initiative's members to collaborate in the advancement of user rights to freedom of expression and privacy. The Principles provide high-level guidance to the ICT industry on how to respect, protect and advance

user rights to freedom of expression and privacy, including when faced with government demands for censorship and disclosure of users' personal information. The Principles are intended to have global application and are grounded in international human rights laws and standards including the Universal Declaration of Human Rights ("UDHR"), the International Covenant on Civil and Political Rights ("ICCPR") and the International Covenant on Economic, Social and Cultural Rights ("ICESCR"). (Global Network Initiative, 2012)

In practice, this is "self-regulation lite." Companies can comply with government censorship demands yet remain in compliance with the GNI's principles if they offer transparency and disclose how they are working within government demands. Yet, to date, new members have not joined the initiative. Other major forces in the social media environment have been noticeably absent. Twitter has not joined "despite their large audience and wide use by activists, in the Middle East and elsewhere" (Kopytoff, 2011). Facebook agreed to join GNI under a newly created observer status that permits corporations considering membership 12 months before making a decision on full membership to participate in policy discussions with members and staff (Feinberg, 2012). (Both Facebook and Twitter are blocked or almost completely blocked in China as well as in other countries including Libya and Vietnam.)

Media theorist Marshall McLuhan once said, "I don't know who discovered water, but it wasn't a fish." With this in mind, we wrote this on a day we abstained from posting, friending, linking, liking, or tweeting a soul. But did anyone notice or care?

## Notes

1. The content-neutrality test has not developed without its problems. As Chemerinsky (2000) notes, the "problem with the Court's application of the principle of content neutrality has been its willingness to find clearly content-based laws to be content neutral because they are motivated by a permissible content-neutral purpose." The court has at times treated content-based laws as content neutral because it said that the law was motivated by a desire to control the secondary effects. He argues the "test of whether a law is content based or content neutral depends not on its terms, but rather on its justification. A law that is justified in content-neutral terms is deemed content neutral even if it is content based on its face."

2. The distinction among different types of communication protected by the Constitution from government regulation is illustrated by obscene and merely pornographic or indecent material. Obscene material has been deemed as unworthy of any First Amendment protection. Pornographic and indecent speech, on the other hand, is constitutionally protected from government regulation. Any regulation of the latter type of communication is subject to a strict scrutiny test.

3. According to the Open Net Initiative, examples of governments filtering or banning YouTube include the following:

~ Thailand: authorities routinely block videos that contain negative messages about the Thai king.

~ Germany, England, United States: authorities have temporarily blocked or removed videos from YouTube due to copyright infringement charges.

~ Mexico: the Federal Electoral Institute ordered YouTube to remove a video defaming incumbent senatorial candidate Fidel Herrera during his re-election campaign.

~ Egypt: authorities required YouTube to block videos and account access for pro-democracy activist Wael Abbas in 2007. His account access has since been restored, but the videos remain blocked.

Some governments block sites during specific events they perceive as a threat to government power or at a particular time. So, for example, Turkish authorities blocked YouTube in 2007 in response to several videos that insulted Mustafa Ataturk and "Turkishness" (http://opennet.net/youtube-censored-a-recent-history).

# References

*Arkansas Writers' Project, Inc. v. Ragland*, 481 U.S. 221, 231 (1987).

Chemerinsky, E. (2000). Content neutrality as a central problem of freedom of speech: Problems in the Supreme Court's application. *Southern California Law Review 74*, 49–64.

Downes, L. (2011, March 30). Why no one will join the Global Network Initiative. *Forbes*. Retrieved from http://www.forbes.com/sites/larrydownes/2011/03/30/why-no-one-will-join-the-global-network-initiative/

E.U. proposes 'right to be forgotten' by internet firms. (2012, January 23). *BBC News Technology*. Retrieved from http://www.bbc.co.uk/news/technology-16677370

Feinberg, A. (2012, May 3). Facebook joins Global Network Initiative as an observer. *The Hill*. Retrieved from http://thehill.com/blogs/hillicon-valley/technology/225217-facebook-joins-global-network-initiative-as-an-observer

Gates, B. (1996, January 3). Content is king. Retrieved from http://www.craigbailey.net/content-is-king-by-bill-gates/

Global Network Initiative. (2012). Retrieved from http://globalnetworkinitiative.org/about/index.php

How to comply with the new E.U. cookie law. (2012). *ComputerWeekly.com*. Retrieved from http://www.computerweekly.com/guides/How-to-comply-with-the-EU-cookie-law

Katz, I. (2012, April 15). Web freedom faces greatest threat ever, warns Google's Sergey Brin. *Guardian*. Retrieved from http://www.guardian.co.uk/technology/2012/apr/15/web-freedom-threat-google-brin

Kopytoff, V. (2011, March 6). Sites like Twitter absent from free speech pact. *New York Times*, p. B4.

McCarthy, C. (2009, February 18). Facebook faces furor over content rights. *CNN*. Retrieved from http://www.cnn.com/2009/TECH/02/17/facebook.terms.service

*Miller v. California*, 413 U.S. 15 (1973).

Pfanner, E. (2012, February 29). France says Google privacy plan likely violates E.U. privacy law. *New York Times*, p. B9.

*Roth v. U.S.*, 354 U.S. 476 (1957).

*Simon & Schuster, Inc. v. New York*, 502 U.S. 105, 118 (1991).

# Index

# Case Index

# Contributors

**Eric Allison**, Ph.D., AICP teaches historic preservation at Pratt Institute in New York City. His latest book, *Historic Preservation and the Livable City*, was published by John Wiley & Sons in January 2012.

**Mary Ann Allison**, Ph.D., is an Associate Professor in the Journalism, Media Studies and Public Relations department at Hofstra University. She is an interdisciplinary scholar who uses media theory, sociology, and complex systems theory to study the ways in which individuals, communities, and institutions are changing. She teaches Media Studies in the School of Communication and is writing a book on new media and society.

**Suzanne Berman** is an Associate Professor in the Journalism, Media Studies and Public Relations department at Hofstra University. She has been a public relations professional for over 20 years working for multi-national corporations including AT&T where she held a number of positions including International Media Relations Director. More recently she has focused her public relations activities on the nonprofit sector directing the public relations efforts of Central Park and working with other nonprofit groups helping them with issues of diversity outreach and other multicultural programs.

**Juliet Dee** is an Associate professor in the Department of Communication at the University of Delaware. She teaches courses in First Amendment law, mass communication and culture, broadcast news writing and television production. She has been director of the Legal Studies Program at the University of Delaware, and has been an editor of the *Free Speech Yearbook*, and is a co-author of *Mass Communication Law in a Nutshell* (2007). She received her bachelor's degree from Princeton University, her master's degree from Northwestern University, and her doctorate from Temple University. She has published articles or book chapters on First Amendment issues involving controversial artwork funded by the National Endowment for the Arts, copyright infringement, anonymous defamation on the Internet, objectionable lyrics in rock music and hip-hop, media liability for violent content, media liability for classified ads for hitmen, and the issue of cyberbullying and the First Amendment.

**Thomas R. Flynn** is a Professor of Communication Law at Slippery Rock University. He has chaired the Department of Communication since 2008. In 2010, he was named a Distinguished Teaching Fellow by the Eastern Communication Association, in acknowledgement of an outstanding career of teaching and educational service. In 2007, he received the Neil Postman Mentoring Award from the New York State Communication Association. In 2004, he received the Eastern Communication Association's Past President's Award for Outstanding Contributions in Scholarship & Service. He has been nominated for the National Communication Association's Franklyn Haiman Award for Outstanding Scholarship in Freedom of Expression. He has also received awards for scholarship from the European Society for Cybernetics and Systems Research, the Kenneth Burke Society, the National Communication Association, and the Eastern Communication Association. His research has been cited in over seventy news stories in print, radio and television, including the *New York Times*, *Roll Call Magazine*, *The Guardian*, and National Public Radio.

**Susan J. Drucker**, J.D., St. John's University School of Law, is a Professor in the Department of Journalism, Media Studies, and Public Relations, School of Communication, Hofstra University. She is an attorney, and treasurer of the Urban Communication Foundation and past president of the Eastern Communication Association and the New York State Communication Association. She is the Communication and Law series editor for Peter Lang Publishing. She has served as editor of the *Free Speech Yearbook*, *Qualitative*

*Research Reports in Communication* (Taylor & Francis) and served as Series editor of the Communication and Law series for Hampton Press. She is the author and editor of 10 books and over 100 articles and book chapters including two volumes of the *Urban Communication Reader*, *Communicative Cites in the 21st Century* (Lang, 2013, in press), *Regulating Convergence* (Lang, 2010), *Voices in the Street: Gender, Media and Public Space*, two editions of *Real Law @ Virtual Space: The Regulation of Cyberspace* (1999, 2005) and *Regulating Social Media: Legal and Ethical Considerations* (Lang, 2013, in press) with Gary Gumpert. She is the recipient of the Franklyn S. Haiman Award for distinguished scholarship in freedom of expression (NCA) and the Donald H. Ecroyd & Caroline Drummond Ecroyd Teaching Excellence Award, Eastern Communication Association, 2012. Her work examines the relationship between media technology, human factors, and city life, particularly as viewed from a legal perspective.

**Bruce E. Drushel** is an Associate professor in the Department of Media, Journalism, & Film at Miami University and directs its Film Studies program. He chairs the Gay, Lesbian & Queer Studies interest group for Popular Culture Association/American Culture Association and has received its Charles Sokol Award and a President's Award for his service to the organization. He is editor of the book Fan Phenomenon: Star Trek and co-editor of the books Queer Identities/Political Realities and Ethics of Emerging Media. His work also has appeared in Journal of Homosexuality, Journal of Media Economics, European Financial Journal and FemSpec, and in books addressing free speech and social networks, free speech and 9/11, media in the Caribbean, C-SPAN as a pedagogical tool, LGBT persons and online media, minority sexualities and non-western cultures, and AIDS and popular culture. He has edited a special issue of *Journal of Homosexuality*, has co-edited a special issue of the journal *Reconstruction*, and shortly will begin co-editing a special issue of *Journal of American Culture.*

**Kelly Fincham** is an Assistant professor in the Department of Journalism, Media Studies and Public Relations at Hofstra University where she teaches advanced online journalism in the undergraduate and graduate program and advises students at the award-winning *Long Island Report* student news website which she founded in 2010. She has more than 25 years of senior editorial experience in both print and online media and has created websites and social media strategies for the Irish Lobby for Immigration Reform, the Irish charity GOAL and the original web space for *The Irish Voice* and *Irish America* magazine, all based in NYC. During her newspaper career she worked at international titles including the *Australian Financial Review* (Fairfax) and *Sydney Telegraph* (News International) in Sydney, and the *Sunday Independent, Irish Independent* and *Evening Herald* (all INM) in Ireland.Her research agenda explores the intersection of social media in journalism practice and curriculum and she is a regular presenter at several national and regional journalism conferences in the U.S. Canada, Ireland and Scotland.

**Gary Gumpert**, Ph.D, Wayne State University, is Emeritus Professor of Communication at Queens College of the City University of New York and President of the Urban Communication Foundation. His creative career as a television director and academic career as a scholar spans over 60 years. In 1960 he directed the Gutenberg Galaxy in which Marshall McLuhan articulated the premise of his forthcoming book. He has authored and edited books include *Talking Tombstones and Other Tales of the Media Age* (Oxford University Press), *The Urban Communication Reader* (Hampton Press) and *Regulating Convergence, Regulating Social Media: Legal and Ethical Considerations* (Peter Lang). He is a recipient of the Franklyn S. Haiman Award for distinguished scholarship in freedom of expression (NCA), the Louis Forsdale Award for Outstanding Educator in the Field of Media Ecology (MEA), and in 2011 received The Neil Postman Award for Career Achievement in Public Intellectual Activity. His primary research and theory agenda focuses on the impact of communication technology upon social and urban space.

**Adrienne E. Hacker Daniels**, Ph.D. University of Wisconsin-Madison, 1993), teaches in the Department of Communication and Rhetorical Studies at Illinois College in Jacksonville, Illinois. She teaches courses in Public Speaking, Freedom of Expression, Mass Communication, Communication Ethics, Rhetorical Theory, and electives including The Rhetoric of Comedy. Her research includes the study of the relationship between rhetorical and poetic theory, and the study of the rhetorical dimensions of more aesthetically grounded artifacts. Hacker Daniels has published essays on Thornton Wilder, Gertrude Stein, W.H. Auden and Hallie Flanagan.

**Dale A. Herbeck** is chair of the Communication Department Studies Department at Northeastern University, where he teaches courses on argumentation, communication law, cyberlaw, and freedom of expression. He is co-author of *Freedom of Speech in the United States*, a past editor of *Free Speech Yearbook*, and a former chair of the National Communication Association's

Commission on Freedom of Expression. His scholarship—which includes numerous book chapters, articles, and papers presented before scholarly societies—has been recognized with the Franklyn S. Haiman Award of the National Communication Association, the James Madison Award from the Southern States Communication Association, and the Robert M. O'Neill Award from the Commission on Freedom of Expression. He has also served as president of the American Forensic Association, chair of the Commission on Freedom of Expression of the National Communication Association, and editor of both *Argumentation and Advocacy* and the *Free Speech Yearbook*.

**Star Muir** is an Associate Professor of Communication at George Mason University, and is currently the Director of Learning Support Services in the Division of Instructional and Technology Support Services (DoIT). Star received his Ph.D. in Rhetoric and Communication from the University of Pittsburgh. His research focuses on environmental communication, and he has published in *ETC.: A Review of General Semantics, Philosophy and Rhetoric*, and the *Speech Communication Teacher*. He is co-editor of *Earthtalk: Communication Empowerment for Environmental Action*. His work on distance delivery of video modules won several Communicator and Telly awards for educational programming. As the current Director of the Student Technology Assistance and Resources (STAR) Center, the Instructional Resource Center (IRC), the Mason Media Lab (MML) and the Center for Training in Office and Productivity Skills (TOPS) at GMU, he is responsible for providing students, staff and faculty with information technology resources and training to accomplish their career and educational goals.

**Warren Sandmann** is currently serving as Provost and Vice President for Academic Affairs at William Paterson University of New Jersey. He has also served in various administrative and teaching positions at Minnesota State University, Mankato, San Jose State University, and the State University of New York at Geneseo. Dr. Sandmann has taught, presented and published on argumentation and debate, communication education, legal studies, communication technology, national sovereignty, free speech decisions of the Supreme Court, and communication law. He also currently serves on the editorial board of the *Free Speech Yearbook*.

**Douglas Strahler** is a faculty member in the Department of Communication at Slippery Rock University of Pennsylvania. He teaches Social Media, Interactive Multimedia and Issues in Communication Technology courses in the undergraduate programs of the department. Douglas earned his Bachelor of Science degree in Communication with an emphasis in emerging technology & multimedia from Slippery Rock University in 2005 and his Masters of Science degree in New Media from the S.I. Newhouse School of Public Communication at Syracuse University in 2006. He is currently pursuing his Doctor of Education degree in Instructional Technology & Leadership from Duquesne University, where he is researching topics dealing with social media in education. Douglas has presented at the Eastern Communication Association (ECA) Convention and New York State Communication Association (NYSCA) Conference.

SUSAN DRUCKER
*Series Editor*

Acknowledging the variety of ways in which the disciplines of communication and law converge, the aim of this series is to publish books at the nexus of these two areas with particular attention paid to communication in law in the changing media landscape.

Utilizing both qualitative and quantitative methodologies, volumes in this series provide analysis of issues at the interdisciplinary and international level such as free and responsible speech, media law, regulation and policy, press freedoms and governance of new media.

For additional information about this series or for the submission of manuscripts, please contact Dr. Drucker at susan.j.drucker@hofstra.edu.

To order other books in this series, please contact our Customer Service Department:

(800) 770-LANG (within the U.S.)
(212) 647-7706 (outside the U.S.)
(212) 647-7707 FAX

Or browse online by series at www.peterlang.com.